Temporary Appropriation in Cities

Alessandro Melis · Jose Antonio Lara-Hernandez ·
James Thompson
Editors

Temporary Appropriation in Cities

Human Spatialisation in Public Spaces and Community Resilience

Springer

Editors
Alessandro Melis
School of Architecture
University of Portsmouth
Portsmouth, Hampshire, UK

Jose Antonio Lara-Hernandez
School of Architecture
University of Portsmouth
Portsmouth, Hampshire, UK

James Thompson
School of Design
University of Melbourne
Melbourne, VIC, Australia

ISBN 978-3-030-32119-2 ISBN 978-3-030-32120-8 (eBook)
https://doi.org/10.1007/978-3-030-32120-8

This Springer imprint is published by the registered company Springer Nature Switzerland AG
The registered company address is: Gewerbestrasse 11, 6330 Cham, Switzerland

Foreword

The Changing Social and Spatial Character of Public Space and Its *Modus Operandi*

Cities are many things. At their best, they offer essential breathing spaces that facilitate social inclusion and enable a diverse range of informal activities. Social inclusion is important, as it is the process by which efforts are made to ensure equal opportunities—that everyone, regardless of their background, can achieve their full potential in life. Gentrification and growing privatization of space are constant threats to this potential. Efforts include policies and actions that promote equal access to public services and space, as well as enable citizen's participation in the decision-making processes that affect their lives in cities. The following chapters presented in this book look at the ways in which 'marginal groups' appropriate these informal spaces. The characteristics of informal, interstitial spaces within the contemporary city, and the numerous creative ways in which they are appropriated, challenge the prevalent critical discourse about our understanding of 'public space', place-making and the character of social order in the city in relation to these informal spaces. These non-planned spaces punctuate the homogenous, controlled, 'official' pattern of public space and the everyday, ubiquitous spaces of the contemporary city. However, they are often overlooked or relegated as 'leftover space' or 'urban voids', excluded from the 'ideal', as they run contrary to the dominant image of the ordered city, such as promoted in common master plans and urban marketing (Sennett, 2018).

Recognizing that cities are places where both problems emerge and solutions are found, at the citizen level, urban activism has been increasingly considered as an important part of urban governance processes, as local actors participate in urban politics and influence decision-making through legal tools and participatory practices.

According to Lewis Mumford, social and spatial inequalities are an intertwined phenomenon and to understand them have to be addressed simultaneously. Chouinard calls them 'places of shadow citizenship and entitlements' (1997) in which inequality and injustice proliferate. This book aims to shed light on the dynamics generating the urban areas and conditions of urban informality and 'shadow citizenship', and to contribute towards understanding the drivers of urban inequalities that pose significant challenges to the existence of urban societies. The more unequal a society is, the more the likelihood of its social polarization to intensify; the wider the spatial and social inequality, the lower the economic growth of a country and consequently the higher the threat to the sustainability of its welfare state (Stiglitz, 2013).

Thus, engendering the dynamics of the United Nation's Agenda 2030's Sustainable Development Goals (Goal 11 calling for 'cities and human settlements' to be 'inclusive, safe, resilient and sustainable'), spatial inequality, and in particular inequalities exacerbated by the economic crisis and austerity measures, has been addressed by activists by means of political protest and by social innovations (Rivera, 2013). Gentrification, growing privatization and the lack of affordable housing as well as accessibility to public spaces have exacerbated the urban conditions of inequality.

How urban informality, public space, growth and inequality are interconnected has long been a field of scholarly reflection and research. It has been argued that cities exist because people tend to agglomerate around space for social, economic or political reasons. Increased job access, housing and amenities, combined with safety and fewer restrictions of personal choices, indicate some critical advantages of urban spaces. Cities form when people try to take advantage of shared infrastructure and cultural amenities. Companies and workers benefit from knowledge creation, job opportunities and matching, as well as information sharing. Agglomeration forces reinforce centralization trends, resulting in urbanization and spatial distribution of location choices and investments, in terms of infrastructure and innovation activities, within cities. The importance of public space lies in understanding it as a manifestation of social relations and a social relation itself (Florida and Mellander, 2018). Even though accelerating urbanization has a significant impact on the spatial outcomes of economic activity and inequalities, it affects in contradictory ways social inclusion in urban areas. Cities in the developed world experience a paradox where rising levels of human capital, economic activity and employment are accompanied by a trend of increasing levels of socio-economic inequality and spatial segregation. Agglomeration effects in cities can act as engines of urban growth, but also as drivers of inequality, not only in economic terms, but also in a wide range of quality-of-life aspects (Lehmann, 2019). Increasing inequalities have become a multi-scale issue. It becomes evident that the urban environments of developed cities face a situation, where despite their continuously rising shares of economic activity, employment and wealth, there are increasing levels of socio-economic inequality, which have a spatial footprint in cities, as well as between neighbourhoods.

Global cities act as both locus and regulator of the uneven processes of capitalism (Harvey 1989), where increasing investment in the financial sector rather than other productive activities has shaped the geography of unequal development. However, the growth of spatial inequalities is not limited to global cities.

Geographical inequalities are a reflection of an out of equilibrium economy, characterized by the dynamic coexistence of regionalized growth and localized decline, by unequal interactions and asymmetrical power relations between cores and peripheries.

The impact of the 2008 economic crisis and subsequent austerity measures accompanied by the persistent stagnation of some economies, along with rapid technological changes in the world of communication and work, have caused conditions of inequality to further proliferate. Alongside regions and cities that have functioned as poles of attraction for private capital investment and innovation, boosted by ad hoc public and private interventions, there are areas in which economic hardship has been coupled with not only private sector but also public sector retrenchment. Therefore, opportunities for social and economic development generated by global economic, financial and political trends have not benefited all countries and cities equally nor have these opportunities been distributed evenly across regions, cities and neighbourhoods, nor among people (Sassen, 1991; Marcuse et al. 2009).

The economic crisis and the ways governments have sought to address it, either with austerity measures or with growth packages, have contributed towards strengthening long-established patterns of unequal territorial and spatial development. Places more equipped with human, infrastructural and social resources have greatly benefited and have had further opportunities to overcome the crisis, while areas with a weaker combination of such factors keep lagging behind. In this scenario, old divisions such as centrality versus periphery in terms of global, regional or urban economic and political centres have exacerbated pre-existing territorial inequality patterns. In Europe, for instance, these inequality patterns are found between Southern and Northern Italy, or West and East Germany, or Northern and Southern England, where gentrification and privatization have created gated communities and immigrant ghettos.

To reclaim the urban space that privatization and gentrification have conquered, it lies in the sphere of civil society where the urban problems are being defined, articulated and dealt with. However, the juxtaposition between spatial inequality and social stratification has dramatically constrained opportunities for specific categories of individuals, with those experiencing a combination of spatial and social disadvantages (e.g. elderly or disabled people in need of healthcare services), finding themselves entrapped in a spiral of deprivation. This spiral of deprivation reveals the interlocking systems of power that have an impact on the most marginalized in society, resonating with Crenshaw's (1989) 'intersectionality of structural identities' based on categories such as gender, race, sexuality and class that interact with each other in various ways and create inequalities and discrimination. In this context, the significance of opportunities for appropriation of public space becomes obvious.

Increasingly, research has focused on the localized clustering of economic activity and the importance of informal entrepreneurialism. Massey's (1984) work on labour policies and consequential spatial division of labour indicates that each stage of economic growth creates new spatial divisions of labour where each period reflects different socio-spatial configurations, introducing new dimensions of inequalities and relations of dominance and dependence between activities in different places and people.

The book *Temporary Appropriation in Cities* conceptualizes and illustrates temporary appropriation of public space as a phenomenon that contributes to the conformation of relevant themes within the urban agenda. The chapters deal with a wide range of the above-mentioned topics, such as citizenship, inequality, social inclusion, migrants, urban social sustainability, public health care, employment, growth, informal entrepreneurialism, urban informality and resilience. Temporary appropriation has behavioural, economic and cultural dimensions, and forms of appropriation are not always obvious: they are not mentioned in building codes and can be subversive and unexpected, emerging in the grey area between legal and illegal activities in the city.

The three editors of this book—Melis, Lara-Hernandez and Thompson—have compiled a highly relevant book that presents a compelling case of the importance of this so far widely overlooked topic that requires further detailed study in the years to come. The 18 chapters of the book examine and map temporary appropriation of public space and the use of public space, in locations as diverse as Auckland, Algiers, Mexico City, Shanghai, Hanoi, San Diego, Tirana and Nairobi.

Offering a better understanding of the appropriation phenomenon, how we should design our public spaces to better facilitate the economic life and social inclusion in cities, the book finally poses the question: How could our streetscapes and plazas better cope with these new complex forms of temporary appropriation, and increasing pressures from homelessness and urban nomadism? The arguments of the authors both develop the concept of temporary appropriation and question it, as well as suggesting alternatives for future expansion.

Steffen Lehmann
University of Nevada, Las Vegas
Las Vegas, NV, USA

References

1. Chouinard, V. (1997). Structure and agency: Contested concepts in human geography, *Canadian Geographer, 41*(4), 363–377.
2. Crenshaw, K. (1989). Demarginalizing the intersection of space and sex. The University of Chicago Legal Forum, 139–167.
3. Florida, R. & Mellander, C. (2018). The geography of economic segregation. *Social Sciences, 7*(8), 123.
4. Harvey, D. (1989). From managerialism to entrepreneurialism: The transformation in urban governance in late capitalism. Geografiska Annaler: Series B, *Human Geography, 71*(1), 3–17.
5. Lehmann, S. (2019). Urban regeneration. A Manifesto for transforming UK Cities in the Age of Climate Change, London: Palgrave Macmillan.

6. Massey, D. (1984). Spatial divisions of labour. London: Macmillan.
7. Marcuse P., Connolly J., Novy J., Olivo I., Potter C. & Steil J. (eds) (2009). Searching for the Just City. Debates in urban Theory and Practice, NewYork/London: Routledge.
8. Rivera, M. (2013). Political criteria for sustainable development goal (SDG) selection and the role of the urban dimension. *Sustainability, 5*(12), 5034–5051.
9. Sassen S. (1991). The Global City: New York, London, Tokyo. Princeton University Press, Princeton, NJ.
10. Sennett, R. (2018). Building and dwelling: Ethics for the City, New York: Allen Lane Publisher.
11. Stiglitz, J. E. (2013). The measurement of wealth: Recessions, sustainability and inequality. International Economic Association Book Series, Contemporary Issues in Macroeconomics, 63–76.

Preface

The book you have in your hands is about temporary appropriation (TA) of the public space which is an urban phenomenon commonly overlooked until now. I have had the opportunity to travel and experience different urban landscapes in cities across the world, in terms of both urban design and how public spaces are used by people.

There is a vast and diverse activity occurring in such public spaces across cultures and environments. However, some of these activities belong to temporary appropriation that as we shall see in the following pages helps to edifice the social landscape of the city, serving to construct the urban identity, social sustainability, public health, crime and the culture of a place. As cities and urban population grow, the understanding of these urban dynamics plays a crucial role to achieve a sustainable, inclusive and healthy urban environment.

This edited book is the result of more than four years of research. It includes contributions from authors whose research is focused on different urban environments across the globe such as Auckland, Hong Kong, Shanghai, Italy, UK, USA and Mexico, illustrating the complexity and multiplexity of TA as a global urban phenomenon. Through our journey, we realized that TA has a multi-scalar and multidimensional nature, meaning that it is highly related to different landscapes within the city such as economic, legal, cultural and political. This compilation aims to illustrate some of these intertwined examples in the contemporary city, highlighting the relevance of the liaison between the design of the built environment and the social landscape in the city.

As academicians and professional urban designers, we are sure that the work is timely and potentially appealing due to its geographical and thematic diversity. Through a series of essays, Temporary Appropriation of the City: A Contribution to Urban Sustainability addresses the human spatialization of public spaces, one of the most important events occurring within the urban landscape. The spatialization of the human being unfolds striking relationships with urban design and planning by using urban spaces as the main stage but varies across countries and cultures.

Newcastle, UK — Jose Antonio Lara-Hernandez
June 2019

Acknowledgements

The work presented here is the result of the support that we have received in three areas: institutional, academic and personal. We would like to thank the University of Portsmouth (UK) and Consejo Nacional de Ciencia y Tecnologia (Mexico) for providing support to carry on the research. We would like to thank our editors, our copy editors, publicists and cover designers. We acknowledge also experts and colleagues who helped with research and gatekeepers who have granted access to research opportunities. We express our gratitude to our partners and friends who have encouraged us to finish this accomplishment. We thank our teachers, students and librarians who have been champions of our work. We acknowledge all the anonymous heroes that help us during the field trip, providing locations, advice, food and water, shelter and even protection. Thank you all.

Contents

Contributors

Thomas Auer Technische Universität München, Munich, Germany

Paola Boarin School of Architecture and Planning, University of Auckland, Auckland, New Zealand

Luisa Bravo City Space Architecture, Bologna, Italy

Ata Chokhachian Technische Universität München, Munich, Germany

Claire M. Coulter University of Portsmouth, Portsmouth, UK

Sean M. Crotty Texas Christian University, Fort Worth, USA

Alexander Gomez-Torres University of Auckland, Auckland, New Zealand

Keith Harris University of Washington, Washington, USA

Brendan C. Josey Monash University, Melbourne, Australia

Alessandra Lai University of Sassari, Sassari, Italy

Jose Antonio Lara-Hernandez University of Portsmouth, Portsmouth, UK

Steffen Lehmann University of Las Vegas Nevada, Las Vegas, USA

Alessandro Melis University of Portsmouth, Portsmouth, UK

Diego Ramirez-Lovering Monash University, Melbourne, Australia

Jeff Rose University of Utah, Salt Lake City, USA

Daniele Santucci Technische Universität München, Munich, Germany

James Thompson University of Melbourne, Melbourne, Australia

List of Figures

Understanding Temporary Appropriation and Social Sustainability

Between Assemblages and Temporary Appropriation: The Case of Mexico City

Temporary Appropriation and Informality

Ongoing Appropriation: Invisible Seattle and Red May

Temporary Appropriation and Public Space: Assessing the CPTED Principle of Activity Support

Temporary Appropriation of Public Spaces: The Influence of Outdoor Comfort

Origins of Informality: Examining the Historical and Spatial Roots of Informal Day-Labor Hiring Sites

Temporary Appropriation in Shanghai and Hong Kong: Two Study Cases Assessing the Resilience of Women Faced with the Lack of Affordable Housing

(Temporary) Appropriation (of Space), Makassar, and Urban Kampung

Extending Temporary Appropriation Through Architecture: The Role of Adaptive Reuse in Shaping New Zealand's Built Environment

Using the Street in Mexico City Centre: Temporary Appropriation of Public Space Versus Legislation Governing Street Use

List of Tables

Using the Street in Mexico City Centre: Temporary Appropriation of Public Space Versus Legislation Governing Street Use

General Introduction

Jose Antonio Lara-Hernandez

Abstract This book conceptualises and illustrates Temporary Appropriation (TA) as an urban phenomenon, evidencing the complex and the multi-scalar nature of TA in relationship to other dimensions and relevant issues of the urban landscape such as sustainability, informality, culture, inclusivity, streetscape design, homelessness, socio-spatial inclusion, building re-use and regulations controlling the use of public spaces. Some forms of appropriation can also be subversive, existing in a grey area between legal and illegal activities in the city. Since urban populations are certain to grow further, this is a key topic for understanding urban dynamics. The book intends to focus on the TA instead as a consequence of the necessity of adapting human needs of a city that deprives the population of reference points due to sudden and unexpected changes. In this last case, the temporality of appropriation becomes an essential element of flexibility, the study of which, however, still constitutes a research gap. Hence, the present book takes into consideration the assumption that the current conjuncture crisis due to the unstoppable growth of the cities imposes a reflection on their development in terms of community resilience, and therefore on the adaptation of the concept of appropriation with respect to the aforementioned phenomena of growth and change.

Keywords Temporary appropriation · Urban design · Public space · Resilience · Social sustainability · Assemblage theory · Informal economy · Informality · Streetscape design · Activity support · Crime prevention · Homelessness · Outdoor comfort · Urban identity · Urban metabolism · Adaptive reuse

The work presented here in a broad sense is about the interaction between people and places within the urban environment. It aims to elucidate the temporary appropriation (TA) as an urban phenomenon, evidencing the complex and the multi-scalar nature of TA in relationship to other dimensions and relevant issues of the urban landscape such as sustainability, informality, culture, inclusivity, streetscape design, homelessness, socio-spatial inclusion, building re-use and regulations controlling the use of public spaces.

J. A. Lara-Hernandez (✉)
University of Portsmouth, Portsmouth, UK
e-mail: Jose.LaraHernandez@myport.ac.uk

A. Melis et al. (eds.), *Temporary Appropriation in Cities*,
https://doi.org/10.1007/978-3-030-32120-8_1

The book intends to focus on the TA instead as a consequence of the necessity of adapting human needs of a city that deprives the population of reference points due to sudden and unexpected changes. In this last case, the temporality of appropriation becomes an essential element of flexibility, the study of which, however, still constitutes a research gap. Hence, the present book takes into consideration the assumption that the current conjuncture crisis due to the unstoppable growth of the cities imposes a reflection on their development in terms of sustainability (Lehmann, 2015), and therefore on the adaptation of the concept of appropriation with respect to the aforementioned phenomena of growth and change.

The following paragraphs describe the term appropriation in the urban environment.

Appropriation, a Brief Background

The Oxford English Dictionary (2015) defines appropriation as the making of a thing private property. However, within the socio-urban design fields, the use of the term has no consensus and remains vague. For instance, some authors (Araya Diaz, 2016; Blanco, Bosoer, & Apaolaza, 2014; Garcia Ramon, Ortiz, & Prats, 2004; Ramirez-Lovering, 2008) use the term when referring to the illegal or informal use of the public space or territory by analysing their time occupancy, utilised design elements and their daily routes. Nevertheless, there is ambiguity when referring to the appropriation as both uses of the public space, illegal and informal, because it belongs to the citizens, and so does its use, and because for most of the governmental authorities, the informality is usually linked to informal commerce.

The psychologist Korosec-Serfaty introduced the term appropriation in the Proceedings of Strasbourg conference in 1976. As a process similar to humanisation (Graumann, 1976), is conceptualised as temporary phenomenon, involving a dynamic interaction between the individual and its environment (Korosec-Serfaty, 1976). For Graumann (1976) the appropriation is the essential societal defined meanings interiorised by the individual. In other words, TA are the activities occurring in public spaces. Considered by authors such as Purcell (2002), Pol Urrútia (2002) and Yory García (2003) as an inborn human need expressed through activities in public spaces as a crucial part of the urban landscape, establishing a theoretical link between people and places. However, the question of what these activities are is still unanswered.

Stickily, when the public space is being used it is being appropriated, although just for a fraction of time. It could be argued that appropriation and the use of the space are similar concepts because as far as an individual exist, she is appropriating the space. Authors, such as De la Torre (2015), Vidal Moranta and Pol Urrútia (2005), Pol Urrútia (2002) and Pol and Iñiguez (1996) argue that the appropriation plays a crucial role constructing the urban identity through individual and collective activities in the urban landscape, arguing that the users only appropriate public spaces where they feel identified. Thus, implying that the appropriation and the use of space are adjacent concepts, not similar though. Users appropriate a determinate area of a public space on daily basis because they feel comfortable there (Franck & Stevens, 2007). Fonseca Rodriguez (2015) provides a definition with potential uses helping to distinguish between the activities related to the appropriation and the simple use of the space. She defines TA as "the temporary act in which people use public spaces to carry out individual or collective activities other than the purpose that the space was originally designed for".

However, this text starts from the assumption that appropriation of the built environment is an inborn socio-spatial special need of the individual (Graumann, 1976; Yory, 2011). Lefebvre (1971) argues that the concept of appropriation is one of the most remarkable results that centuries of philosophical reflections have bequeathed. He describes appropriation as the goal of the social life by claiming that the domination of the natural environment without the appropriation tends to be absurd. Yory (2011), corroborates this argument by stating that we as humans are connaturally engaged with the construction-appropriation of our environment. Moreover, the concept of appropriation is embodied in his theoretical definition of topophilia as "the act of co-appropriation generated between the man and the world; through which the world becomes the world, at the opening realised by the man within its historic-spatial nature and human becomes human through its spatialisation" (Yory, 2011: p. 45).

Temporary Appropriation in the Urban Environment

Regarding the urban landscape, the appropriation can also be defined as the interaction between citizens and their city expressed through certain kinds of activities occurring in public spaces (Lara-Hernandez, Melis, & Caputo, 2017). These activities contribute to the edifice of the social urban landscape, which is embedded in the second aspect of the right to the city. Purcell (2002), inspired by ideas of Lefebvre, argues that the appropriation allows citizens to produce urban space rather than just to use the already produced urban space. The appropriation is what gives citizens the right to fully use and manage their everyday life within the urban environment (Lefebvre, 1992). Graumann (1976) contributes to reinforce this position by stating that the appropriation of the public space is a medium and a goal in order to overcome human alienation. Thus, it can be inferred that appropriation of space that is ephemeral and time framed could be understood as TA, in which the design of the built environment is necessary but not sufficient (Graumann, 1976). Along the same trajectory, Martínez (2014) emphasises that the TA relies on both the configuration of the built environment and the meanings and implicated relationships (people's values, behaviours and actions) of the social life in the city.

It follows that the TA is an individual, social, and spatial need that cannot be underestimated when it comes to urban studies. Public spaces and their design are strongly associated to the possibility of occurring activities related to TA (Gehl, 2010; Madanipour, 1996; Moudon, 1991; Noschis, Dosio, Feddersen, & Triantis, 1978). Thus, while appropriation is a broader term, its temporary variation refers more specifically to urban public spaces. Although there is conspicuous literature on the more general theme of appropriation, a phenomenon that refers to a lasting and consistent condition over the years (Ardura Urquiaga, 2014; Díaz & Ortiz, 2003; Ramirez-Lovering, 2008) and, even if less, on TA, that refers to non-permanent, daily or extemporaneous appropriation (Fonseca Rodriguez, 2015; Lara-Hernandez et al., 2017). The accent in the latter has always been placed on the informality of this action.

The appropriation of the built environment as a continuous synergy between citizens and the urban landscape displayed through specific activities that contributes to the edifice of the social urban landscape (Lara-Hernandez et al., 2017). When citizens appropriate public spaces, they participate in producing urban space beyond merely inhabiting an urban space already formed and receive the right to manage and use their everyday life completely (Lefebvre, 1992). Relph (1976) defines place as the centre of action, a blending realm between human and natural spheres and as the heart of our prompt experiences of the world. Thus, the space endowed with meaning and values, determined by culture occupied by a person or a thing are places (Madanipour, 1996). Graumann (1976) reinforces that argument by claiming that the appropriation of the public spaces contributes to overcome human alienation. The latter implies that the fleeting and time-framed appropriation might be assumed as temporary appropriation (TA), in which the design of the built environment is crucial but not sufficient (Graumann, 1976). Martínez (2014) support this argument by claiming that TA is dependent upon both the design of the built environment and the cultural connotations of its social realm. For instance, Lehmann (2009) wrote about strategies for informal urban interventions and the influence of site-specific artistic concepts in the creation and appropriation of informal public space. He notes that 'vacant city lots and buildings are often used as starting points for cultural innovations. Over the course of the transformation of the post-industrial city, public space has become an exciting laboratory for interdisciplinary cooperation between artists, architects, urban planners, and landscape architects. Consequently, TA is a spatial, and social need that must be taken into consideration in regards research within the urban field.

In the following chapters it will be argued the multi-scalar and highly complex nature of TA in relation to other relevant urban issues such as sustainability, health, informality, crime, outdoor comfort, theoretical discussions focused on insurgent urban space, crime and urban design, environmental comfort, day-labour, homelessness, women in public spaces, emergent assemblages, citizenship, political ecology, urban culture and identity.

Chapter "Understanding Temporary Appropriation and Social Sustainability" frames within the urban context, the relevance of the temporary appropriation in relation to social sustainability, identifying and categorising temporary appropriation into three main groups of activities featuring the urban landscape. It illustrates how the TA as concept is aligned with what Seghezzo (2009) calls *persons,* as one of the fifth dimensions of sustainability, differing from the definition used by Leary-Owhin (2015), and influenced by Lefebvre, that describes the TA as a time span condition by differential spaces in the urban context.

Chapter "Between Assemblages and Temporary Appropriation: The Case of Mexico City" explores the way in which temporary appropriation could be interpreted as an assemblage product of other assemblages within the urban landscape. It unravels aiming to re-think TA nature through intertwined theoretical frameworks such as assemblage theory. The chapter combines a diverse range of theories conceptualising TA as an essential feature of the urban landscape and as an emerging product of other assemblages such as culture, legal framework and urban design. To do so it illustrates Mexico City Centre as an example of a highly coded city in which these assemblages emerge, taking a representative sample street as a case-study to analyse TA in relation to the streetscape design through participant observation and image analysis of the visual complexity of the streetscape.

Chapter "Temporary Appropriation and Informality" explores the definitions and uses of both informality and TA in an urban context, illustrating TA as an overlooked urban phenomenon and misinterpreted as a form of informal practice. By separating TA from informality, our team provides a simpler model for understanding urban informality, whilst at the same time consolidating TA as a concept in its own right. Existing definitions of informality as applied in studies concerning informal practices in the Global North are challenged, arguing that some understandings of informal behaviour in urban settings could be better understood when viewed as forms of TA. A categorisation of temporary appropriation is provided with examples from different cities and cultures, showing that there is a definite need for a deeper understanding of temporary appropriation and its relationship with the urban informality of any socio-cultural landscape.

In Chapter "Ongoing Appropriation: Invisible Seattle and Red May" Keith Harris illustrates Invisible Seattle and Red May as events that were a mature and explicitly anti-capitalist development of a project from the early 1980s, serving as both a history of the movement between the two groups and their work, as well as a theoretical contribution to the notions of insurgent urban space (Hou, 2010) and nomad citizenship (Holland, 2011). The central argument is that although urbanists and activists often focus on the spatial dimensions of occupation, the temporal dimension is perhaps even more crucial for the constitution of collective subjectivity (Braidotti, 2013).

In Chapter "Temporary Appropriation and Public Space: Assessing the Cpted Principle of Activity Support" Alexander Gomez-Torres assesses the role of activities related to urban design as contributor of TA of the public space for crime prevention purposes. It presents a case study which critically assesses the Third Millennium Park in Bogota DC, Capital District of Colombia. The results unveiled a significant low level of activities and a tendency for using signage mainly as sets of instructions for mandatory purposes rather than a strategy for helping TA of the public space. The research chapter concludes that the urban design intervention played a positive role in the reduction of crime within the areas in the case study. Though, its concentration variously increased throughout the locations in the surrounding areas. Hence, implications of territoriality, crime displacement and the access to comprehensive crime data should also be considered in further research.

In Chapter "Temporary Appropriation of Public Spaces: The Influence of Outdoor Comfort" Daniele Santucci, Ata Chokhachian and Thomas focus on TA using an approach highlighting the necessity of adapting human comfort to the needs of a city to compensate sudden and unexpected changes due to urban microclimate. In the context of the enormous urbanization phenomena occurring all over the world, the relevance of this issue increasingly gains importance. There is still an open question, if the daily public life or use of urban spaces can be acknowledged as appropriation or not. If yes, is this temporary or permanent? The concept of appropriation is closely tied to the sense of identity and comfort as well, besides temporal factors. This chapter will conclude supporting the idea that the temporary appropriation takes place in urban public spaces in which individuals or group of people feels culturally and environmentally identified.

In Chapter "Origins of Informality: Examining the Historical and Spatial Roots of Informal Day-Labor Hiring Sites" Sean Crotty explores the daily-appropriation of space for labour-seeking, drawing on more than five years research in the San Diego Metropolitan Area. He analyses the *places* where hiring sites operate to demonstrate the linkages between neighbourhood characteristics and the establishment of day-labour hiring sites. Ethnographic data further supports an argument that hiring sites exhibit a high-degree of locational "stickiness," due to the historical, economic, and emotional content embedded in each site through the production of space. The chapter concludes by examining the specific ways that laborers and other stakeholders interact to ensure that access to space is maintained and in so doing, foster relationships that produce socially-sustainable outcomes for day-laborers and their neighbours.

In Chapter "Unsheltered Homelessness and the Right To Metabolism: An Urban Political Ecology of Health and Sustainability" Jeff Rose explores the relationship between TA and unsheltered homeless utilising an urban political ecology approach, considering the ways in which a right to metabolism demonstrates that the temporary appropriation of space by those facing homelessness is a biopolitical claim to survival. In this appropriation of public space, those facing unsheltered homelessness lay bare the socioenvironmental relations that characterize aspects of all human life, including eating, breathing, disposing of wastes, and others. These human metabolic processes for people facing unsheltered homelessness uncovers a justice-focused analysis in which a "right to metabolism" is a fundamental necessity for city life.

In Chapter "Temporary Appropriation in Shanghai and Hong Kong: Two Study Cases Assessing the Resilience of Women Faced With the Lack of Affordable Housing" Alessandra Lai discusses the scholastic notion of 'temporary appropriation' of the public space, referring to a wide range of urban scenarios that evoke conditions of either spatial, temporal or socio-economic disorder. Such disorder is manifested by the frequent contrast between the planned purpose of a public space and its actual use. Often, the unforeseen habits of the users of a place obscure the planned design narrative, adding a juxtaposed layer to the urban life. The chapter investigates two phenomena in relation to its occurrence in public spaces in China, the characteristic spreading of various domestic functions (such as cooking and laundry) onto the sidewalks of Shanghai and the widespread use by Filipino maids of the squares and bridges of Hong Kong during their days off. It highlights how such 'user invasions' reflect the resilience of these largely female communities in order to effectively and efficiently face issues related to affordable housing.

In Chapter "(Temporary) Appropriation (of Space), Makassar, and Urban Kampung" Brenda C. Josey and Diego Ramirez-Lovering discuss TA in the context of informal settlements and urban kampungs in Makassar, South Sulawesi, Indonesia. Case study results suggests that emergent and regular instances of TA within the informal context are a manifestation of interactions between legal frameworks, socio-spatial histories and practices, community cohesion, changing land occupancy patterns, and settlement morphologies. The chapter advances that TA is an urban phenomenon with potential to be harnessed for design development; however, simultaneously it postulates that the concept of TA should be evaluated and extended by questioning what is understood as appropriation in complex urban contexts. Further to this, the chapter questions what constitutes TA in circumstances where open or public space is non-existent, limited, or contested. Through such an analysis, it is argued that instances of TA may serve to render visible formerly hidden social practices, needs, desires, and shortcomings of existing development patterns, and thus be leveraged and operationalised to inform the design of community engagement practices and built environment solutions.

In Chapter "Extending Temporary Appropriation Through Architecture: The Role of Adaptive Reuse in Shaping New Zealand's Built Environment" Paola Boarin explores the multiple meanings and implications of integrated building adaptation and its role in the retention and re-creation of heritage values and sense of place. The interdependence between built and natural environment, as well as the building and urban dimension, is discussed to highlight its relationship with adaptive reuse and temporary appropriation. These concepts are further analysed to appreciate how New Zealand's cultural plurality enriches the understanding of cultural heritage and the approaches related to the adaptation of the built environment. It discusses how contemporary architecture in New Zealand is conveying cultural, environmental and social values, especially in the adaptive reuse of existing and historic buildings, in the attempt of establishing a new dialogue between tradition and innovation that could overcome temporality and extend the concept of appropriation of space.

Chapter "Using the Street in Mexico City Centre: Temporary Appropriation of Public Space Versus Legislation Governing Street Use" discusses the function of streets in cities and their relevance towards the construction of the urban landscape, highlighting the role of TA for understanding social dynamics within a given urban environment. It illustrates Mexico City Centre as a case study, examining the laws and regulations set out by the government of Mexico City which regulate the use of the street. It contrasts this with the ways in which the inhabitants of the city appropriate public space on a daily basis. There is a discrepancy between the legislation surrounding potential activities occurring on the street, and a seemingly tacit consensus between citizens regarding how they appropriate such public spaces. The chapter explores this contrast, and blueprints unexpected and sometimes irrational manners in which public space is used in traditional ways, how creative ways are found to use the street area within the spirit of the law, and where further research on this topic this could lead in future.

In Chapter "Transforming Everyday Public Space: Human Appropriations in Search for Citizenship and Urban Well-Being" Luisa Bravo discusses a new concept of urbanity and public life in the contemporary city, in which spontaneous actions are promoted by private groups, redefining the collective sphere. It highlights the work made by urban designers, creating new places for the benefit of communal life, potentializing the temporality to reach the community's imagination, utilising an emotional cornerstone allowing them to take part in the design and making process of their urban landscape.

References

Araya Diaz, D. F. (2016). Propiedad y apropiación. In *Contested Cities*. Madrid, España. p.

Ardura Urquiaga, Á. (2014). Madrid. Espacio público confiscado. La privatización y resignificación del espacio público en los procesos de transformación material de las plazas del centro de Madrid. *Scripta Nova. Revista Electrónica de Geografia y Ciencas Sociales*. [Online]. Available from http://www.ub.edu/geocrit/coloquio2014/AlvaroArduraUrquiaga.pdf.

Blanco, J., Bosoer, L., & Apaolaza, R. (2014). Movilidad, apropiación y uso del territorio: una aproximación a partir del caso de Buenos Aires. *Scripta Nova. Revista electrónica de geografia y ciencias sociales, XVIII*(493).

Braidotti, R. (2013). *Metamorphoses: Towards a materialist theory of becoming*. John Wiley & Sons, Inc.

de la Torre, M. I. (2015). Espacio público y colectivo social. *Nova Scientia, 14*(7), 495–510.

Díaz, F., & Ortiz, A. (2003). *Ciudad e Inmigración: Uso y Apropiación del Espacio Público en Barcelona* (pp. 399–407). [Online]. Available from http://www.insumisos.com/lecturasinsumisas/Apropiaciondelespaciopublico.pdf.

Fonseca Rodriguez, J. M. (2015). La importancia y la apropiación de los espacios públicos en las ciudades. *Revista de Tecnologia y Sociedad, 4*(7), 1–11.

Franck, K. A., & Stevens, Q. (2007). *Loose space: Diversity and possibility in urban life*. New York: Routledge.

Garcia Ramon, M. D., Ortiz, A., & Prats, M. (2004). Urban planning, gender and the use of public space in a peripherial neighbourhood of Barcelona. *Cities, 21*(3), 215–223. [Online]. Available from https://doi.org/10.1016/j.cities.2004.03.006.

Gehl, J. (2010). *Cities for people*. Washington DC: Island press.

Graumann, C. F. (1976). *The concept of appropriation (aneignung) and modes of appropriation of space, 6*(2), 301–313. [Online]. Available from http://iaps.scix.net/cgi-bin/works/Show?iaps_00_1976_009.

Holland, E.W. (2011). Nomad citizenship: Free-market communism and the slow-motion general strike. University of Minnesota Press.

Hou, J. (2010). *Insurgent public space: Guerrilla urbanism and the remaking of contemporary cities*. New York: Routledge.

Korosec-Serfaty, P. (1976). Appropriation of space. In *Proceedings of the Strasbourg Conference. IAPC-3*. 1976 Strasbourg-Lovaine La Neuve: CIACO. p.

Lara-Hernandez, J. A., Melis, A., & Caputo, S. (2017). Understanding spatial configuration and temporary appropriation of the street in Latin American cities: The case of Mexico City Centre. In H. Boudagh, A. Versaci, F. Trapani, & M. Migliore et al. (Eds.), *Advances in science, technology & innovation* (pp. 153–173).

Leary-Owhin, M. (2015). A fresh look at Lefebvre's spatial triad and differential space: A central place in planning theory? In *2nd Planning Theory Conference University of the West of England* (pp. 1–8).

Lefebvre, H. (1971). *De lo rural a lo urbano*. In J. Gonzalez-Pueyo (Ed.), Barcelona, España, Ediciones Peninsula.

Lefebvre, H. (1992). *The production of the space* (1996th ed.). Cambridge: Blackwell Publishers Ltd.
Lehmann, S. (2009). *Back to the city: Strategies for informal urban interventions: Collaboration between artists and architects*. Hatje Cantz.
Lehmann, S. (2015). Low carbon cities: More than just buildings. In S. Lehmann (Ed.), *Low carbon cities: Transforming urban systems* (pp. 1–56).
Madanipour, A. (1996). *Design of urban space. An inquiry into socio-spatial process*. New York: John Wiley & Sons Inc.
Martínez, E. (2014). Configuración urbana, hábitat y apropiación del espacio. *Scripta Nova. Revista electrónica de geografía y ciencias sociales. XVIII*(493), 20.
Moudon, A. V. (1991). *Public streets for public use*. New York: Columbia University Press.
Noschis, K., Dosio, M.-J., Feddersen, P., & Triantis, E. (1978). Appropriation of space: A method and two case studies. *Ekistics, 45*(273), 451–466.
Oxford University Press. (2015). *Oxford english dictionary online*. [Online]. Available from http://dictionary.oed.com. Accessed May 12, 2015.
Pol, E., & Iñiguez, L. (1996). La Apropiacion del Espacio. *Monografías Psicosoaicles Ambientales*, 9.
Pol Urrútia, E. (2002). El modelo dual de la apropiación del espacio en Mira, R. Sabucedo, J. M. y Romay, J. *Psicología y Medio Ambiente. Aspectos psicosociales, educativos y metodológicos* (pp. 123–132).
Purcell, M. (2002). Excavating Lefebvre: The right to the city and its urban politics of the inhabitant. *GeoJournal, 58*, 99–108. [Online]. Available from https://doi.org/10.1023/b:gejo.0000010829.62237.8f.
Ramirez-Lovering, D. (2008). *Opportunistic urbanism*. Melbourne: RMIT Press.
Relph, E. (1976). *Place and placelessness* (2nd ed.). Pion Ltd.
Seghezzo, L. (2009). The five dimensions of sustainability. *Environmental Politics, 18*(4), 539–556. [Online]. Available from https://doi.org/10.1080/09644010903063669.
Vidal Moranta, T., & Pol Urrútia, E. (2005). La apropiación del espacio: Una propuesta teórica para comprender la vinculación entre las personas y los lugares. *Anuario de Psicologia, 36*(3), 281–297. [Online]. Available from https://doi.org/10.1006/jevp.2000.0185.
Yory García, C. M. (2003). *Topofilia, ciudad, territorio: una estrategia pedagógica de desarrollo urbano participtivo con dimensión sustentable para las grandes metrópolis de América Latina en el contexto de la globalización: 'el caso de la ciudad de Bogotá'*. [Online]. Universidad Complutense de Madrid. Available from http://biblioteca.ucm.es/tesis/ghi/ucm-t26725.pdf.
Yory, M. C. (2011). El Concepto de Topofilia entendido como Teoría del Lugar. *Revista Barrio Taller* (pp. 1–17).

Understanding Temporary Appropriation and Social Sustainability

Jose Antonio Lara-Hernandez and Alessandro Melis

Abstract This chapter reviews the relevant literature about temporary appropriation as a concept within the urban landscape . A review of theoretical and empirical texts reveals that despite the lack of a defined theory in this specific field, the temporary appropriation is a key concept for the understanding of people's sense of belonging to a place, and identity. The chapter fills the research gap through an extensive cross-disciplinary literature review that includes topophilia, space production, and environment and behaviour theories. The aim of the present writing is to frame, within the urban context, the crucial role of temporary appropriation in relation to social sustainability. To do so, it identifies and categorises the temporary appropriation into three main groups of activities featuring the urban landscape.

Keywords Appropriation · Social sustainability · Temporary appropriation · Urban sustainability

Introduction

This chapter will argue that there is a strong link between the TA and social sustainability, proving to be a complementary relationship. It will illustrate TA of public spaces as phenomenon that plays a key role in the urban dynamics by contributing to the construction of the social landscape of the city. Moreover, the processes of the social landscape in a city a strongly linked to the resilience of an urban environment (Childers et al., 2014), in which the design of public spaces has implications towards TA. We agree with researchers such as Clark and Dickson (2003) and Kates et al. (2001) to push further the need of studies about human transformation as essential to the science empirical agenda for the common interest. The following section explain both concepts TA and the simple use of the space, elucidating their differences and similarities in terms of urban design.

J. A. Lara-Hernandez (✉) · A. Melis
University of Portsmouth, Portsmouth, UK
e-mail: Jose.LaraHernandez@myport.ac.uk

A. Melis et al. (eds.), *Temporary Appropriation in Cities*,
https://doi.org/10.1007/978-3-030-32120-8_2

Temporary Appropriation Vs Use of the Space

In the General Introduction was highlighted the significance of the TA and its relevance to the urban context by establishing the bond between people and places: through a set of activities occurring in public spaces. However, the question of what these activities are is still pending. If one wished to be pedantic, one might say that we appropriate public space whenever we passed through it, however transitory our appropriation may be. Torres (2009) considers the use and the appropriation of space to be similar concepts by claiming that as far as the individual exists, he is appropriating the space. Other authors (de la Torre, 2015; Pol Urrútia, 2002; Pol & Iñiguez, 1996; Vidal Moranta & Pol Urrútia, 2005) instead suggest that the use and the appropriation of the space are adjacent concepts, not similar though. In their view it is appropriation, in the form of activities whether by one person or by several, that confirms the location's identity; they claim that public spaces are only appropriated by users who feel themselves identified there. This position is supported by studies done by Franck and Stevens (2007), whose interview results illustrate how regular users appropriate a specific area of a public space on a daily basis because they felt it to be comfortable.

A definition of the potential uses is provided by Fonseca Rodriguez (2015) when she describes the TA as "the temporary act in which people use public spaces to carry out individual or collective activities other than the purpose that the space was originally designed for." This definition helps, in fact, to distinguish between the activities related to the appropriation and the simple use of the space. There are similarities here with *persons* in the sense in which they were defined by Seghezzo (2009) as one of the five dimensions of sustainability. Leary-Owhin (2015), who was influenced by Lefebvre, take a different view, describing temporary appropriation as occurring when urban spaces are occupied for a period of time.

Nevertheless, not all the activities that a public space could host are a spatialised expression of the TA. One of the main challenges is in fact to identify the activities taking place in public spaces that are relevant for the phenomenon according to their social, cultural and environmental context-specific nature. Consistently with scholars such as Crossa, Drummond, and Monnet, it is possible to group into three different categories the activities that can be considered as an expression of the TA:

- Activities related to the economy (Crossa 2016, 2009; de la Torre, 2015; Kim, 2013; Lara-Hernandez & Melis, 2018; Loukaitou-Sideris & Ehrenfeucht, 2009). See Fig. 1.
- Activities related to leisure such arts and sports (Díaz Larrañaga, Grassi, & Mainini, 2011; Drummond, 2000; Hernandez Mendo & Morales Sanchez, 2008; MacDonald & Shildrick, 2007; Mouffe, 2014; Simpson, 2011; Seaman & McLaughlin, 2014). See Fig. 2.
- Activities related to sacralisation (Portal, 2009; Monnet, 1995). See Fig. 3.

Table 1 summarises the activities in which people temporary appropriate the public spaces.

Fig. 1 Man selling in streets in Palermo City Centre (left); man selling candies in Mexico City Centre (right) (*Source* Authors)

Fig. 2 (left) Men watching football match in the street in Palermo City Centre; (right) Woman cooking in street in Mexico City Centre (*Source* Authors)

The Relationship Between Temporary Appropriation and Place

Although in theory, every space could be temporarily appropriated. It is not clear, however, which spaces are suitable to be appropriated or why. Within this context, Franck and Stevens (2007) defined what they called "loose space," a condition affected by how people perceive specific public spaces and the possible activities that might be appropriate their. In their view, an urban public space is only loose

Fig. 3 (left) Altar in street in Palermo City Centre; (right) Altar in street in Mexico City Centre (*Source* Authors)

Table 1 Activities in the public space related to temporary appropriation (*Source* Lara-Hernandez et al., 2017)

Category	Economy		Leisure			Sacralisation
Description	Any activity in which a person or group use the public space in order to obtain an economic benefit directly or indirectly		Any activity in which a person or a group use the public space for leisure purposes			Any activity in which a person or a group use the public space for religious purposes
Sub-category	Work	Trade	Sports-games	Artistic expres-sions	Rest	
Individual or collective	Advertising or promoting services, waiting, engaging or attracting possible clients	Selling or buying products (food, handicraft, clothes, etc.)	Skateboarding, soccer, cards, marbles, hopscotch	Playing music, dancing, painting, acrobatics, reciting and singing	Eating, resting	Processions, praying, lighting candles and putting flowers

when its possibilities are realised and serves the purpose to meet the requirements of the people carrying out TA; it is necessary if the space is to be called "loose" that the activities implied by the possibilities are not those for which the space in question was designed or intended. Among loose space's characteristics, the most striking are its diversity, its disorder, and above all its potential. While public space anywhere in the world can be loose, the activities that make it so are not necessarily characteristic of TA, which—it would seem—occurs in [in the view of Relph (1976)] a place that is a centre of action and blends the natural and the human, reflecting how we experience the world. This may be interpreted as saying that place exists when space is occupied, whether by a thing or person, and the occupation is what gives it meaning and cultural value (Madanipour, 1996). This meaning refers to the symbolic and imaginary attributes that people have about a determinate place, building or urban element and what they represent for them (Rapoport, 2005). Moreover, the place is embodied by the cultural landscape in which the individual plays a key role in its conformation (Hubbard, Kitchin, Bartley, & Fuller, 2002; Seghezzo 2009). This view is supported by several authors such as Graumann (1976, 1983), Pol Urrútia (2002), Vidal Moranta and Pol Urrútia (2005) and Martínez (2014), that claim that people temporary appropriate public spaces where they feel identified or somehow represented by. In addition, researchers of environmental psychology, such as Vidal Moranta and Pol Urrútia (2005) hold the view that the appropriated place represents for the self stability and continuity side-by-side with a group's identity and cohesion. TA happens in any public urban space where the individual or group experiences cultural identification.

Temporary Appropriation and Culture

As stated in the previous paragraph, the TA is activated by the sense of belonging whose nature is predominantly cultural. However, the term culture is subject to different interpretations depending on the discipline. The Cambridge English Dictionary (2018) defines culture as "the way of life, especially the general customs and beliefs, of a particular group of people at a particular time". The Oxford English Dictionary (2018) defines culture as: (1) the arts and other manifestation of human intellectual achievement regarded collectively, (2) the ideas, customs, and social behaviour of a particular person or society. Bonfil Batalla (2004) defines culture more broadly, seeing it as a group of symbols and meanings together with knowledge and skills, attitudes and values, social structure, ways of communicating and physical objects that make it possible for a society to live the life it does as well as allowing itself to be transformed and reproduced in generations to come. The built environment is a small subset of that enormous and all-embracing domain we refer to as culture. According to Moudon (1991), because different societies behave in different ways and use space accordingly, the built environment is culturally specific. In the Netherlands, for example, people place strict demarcation lines between their private and public lives and public space is a space for strangers, where the superficial and the spontaneous may be found (Haan, 2005). In Mexico, on the other hand, private/public boundaries are blurred and there is constant temporary appropriation of public space (Monnet, 1995). In the built environment our systems—settings systems, activity systems—and the way the space is designed has a significant influence on these. Rapoport (1998) says that it is possible to break any activity in public space, whether TA or not, into four components. These are: the activity in question; how the activity is conducted; the activity's association with other activities and its integration into an activity system; and the activity's meaning. According to Kyle, Jun, and Absher (2014), the space's identity and how it bonds with individuals is aided by activities that constantly re-occur in the space over a period of time and that, thanks to their more or less harmonious nature, can all be pursued individually and for their own sake. This, in the same author's view, constitutes culture (Eliot, 1949). Thus, we can infer that culture is a dynamic condition in which activities over time attain culture, but these activities also are a reflection of any specific culture.

In this respect, the appropriation relates to the social sustainability as initially defined in the Brundtland report (1987). Hence, within the urban context, urban social sustainability aims to not compromise the social needs that the built environment can satisfy for future generations. Consequently, the aim of this chapter is to illustrate and explain the relationship between these two concepts, the TA and social sustainability on their theoretical and practical basis. The following section illustrate the concept of social sustainability within the urban scenario.

Urban Social Sustainability

The previous paragraphs described the three main groups of activities that characterise the TA in the public spaces as a spatialised expression and described where the TA occurs, and its relationship with the conceptualisation of culture. Following the Brundtland report (1987), several scholars (Dixon et al., 2009; Dempsey, Bramley, Power, & Brown, 2011; Lozano, 2008; Seghezzo, 2009; Woodcraft, 2012) refer to the aforementioned relationship as core aspects of the social and environmental sustainability. The present section will therefore focus on the social sustainability concept within the urban context and illustrates its relationship with the ideas already expressed.

According to Brundtland's three pillars of sustainable development, several authors, such as Costanza (1992), Kleine (2009), Munasinghe (1993) and Spindler (2011) have provided a number of definitions for Environmental, Economic, and Social Sustainability. However, according to Seghezzo (2009), these three heads unduly narrow the perspective by over estimating the power that economic reasoning has to explain things at the cost of sufficient attention to development's essential aspects. In turn Vallance, Perkins, and Dixon (2011) also take the view that economic and environmental aspects should be addressed only after the question of sustainable social development has been satisfactorily dealt with, and they have introduced three categories for discussion that has taken place on the matter of social sustainability:

developmental, bridge and maintenance. Developmental social sustainability has to do with access to such goods as justice and equity, shelter, medicines, and fresh food and water. In an argument that reflects the Maslow hierarchy of needs (Maslow, 1943), the assumption is that the environment will begin to benefit after these fundamental needs been met. The idea behind bridge social sustainability is to identify which social conditions are necessary if ecological sustainability is to be supported, and the conditions may be either transformative or non-transformative. Transformative conditions involve re-examining how people interact with the environment and involve challenges to fundamental ideas about the social construction of the environment, while non-transformative conditions require no change in ideology but simply a better or even simply different way of doing things. Examples would be the use of technology, for example with electric buses. The third category, maintenance social sustainability, has to do with ways in which it is possible to preserve the environment and socio-cultural attributes. The purpose for all three conditions is to create a level of social consciousness about how human action affects the natural environment rather than to produce ideas about making the social and cultural fabric sustainable. Moreover, this arrangement does not recognise the spatial needs of the individual because the human component is represented by society as a whole.

Social sustainability within the urban context aims to not compromise the social needs that the built environment can satisfy for future generations. It combines both social and physical design of an infrastructure that can support both cultural and social life (Woodcraft, 2015). According to Dempsey et al. (2011), social equity and community sustainability are both important if urban social sustainability is to be achieved. The first aspect is frequently associated and measured by the accessibility to key facilities that satisfy basic needs within a determinate area (Barton, 2002; Burton, 2000). It is a common assumption that social equity means that all members of society are included with no distinction is made between them and that social exclusion and urban segregation are mitigated by public round appropriation (Pol, 2002). The second aspect is frequently quoted in relation to community formation and the extent to which people interact socially, since people who live in a particular area and do not interact socially cannot be viewed as a community but simply as people who take little pride in a place and show no attachment to it but simply lived there separate existences (Dempsey, 2006). Studies carried out by Seaman and McLaughlin (2014) indicate that the degree to which individuals interact with their outdoor environments is directly proportional to the amount of attachment they feel to the place and to their sense of community. According to Lofland (1998), the public realm is created in the public space and this is what makes social territory relevant in an urban environment. Both of these arguments allow us to infer that people who either are prevented from interacting with their built environment or simply do not do so do not merit the description of citizens.

Returning briefly to the pillars of sustainability, other researchers such as Deakin, Mitchell, Nijkamp, and Vreeker (2007), Hawkes (2001), Lozano (2008) and Valentin and Spangenberg (2000) have added a fourth pillar. At the heart of this pillar are culture and governance along with well-being and justice (Ali-Toudert & Ji, 2017). Also, Musacchio (2009) put forward the idea of a model of sustainability supported by what she called the six Es: aesthetics, environment, ethics, equity, experience and economy. These models of sustainability can be used to represent neighbourhood, city and regional urban scales as all of them recognise the social realm's importance, but they fail because they regard the social realm as a matter of the whole of society and not as the set of individuals that is public space and none of them includes a consideration of time, even though sustainability is not static but dynamic (Antrop, 2005). The proposals by Seghezzo (2009) are therefore more accurate because they embrace "place, permanence and persons" where place is not only geographical and physical space in three dimensions but also the cultural landscape while permanence has to do with time and persons are society's individuals. Figure 4 illustrates the model already described.

Urban social sustainability can therefore be seen as an umbrella expression embracing cities' environmental, social and economic realms. Social sustainability allows the organisation of cities and societies in a way that meets the needs both of the current and future generations while not involving and ecological compromise (Vojnovic, 2014), which, as previously discussed, addresses the design needs of both the social and the physical realms (Woodcraft, 2015) which Capel Sáez (2002) suggests are both urban landscape elements. This argument is supported by Antrop (2005) who stresses that cultural, architectural and other landscapes overlap in the urban landscape and are constantly in flux as they dynamically express the interaction between the environment's cultural and natural forces. Musacchio (2009) describes the urban landscape as a system constantly re-shaped as people's needs change and claims that people's actions, behaviours and values influence in designed landscapes' heterogeneity both functionally and spatially and that this is a form of appropriation.

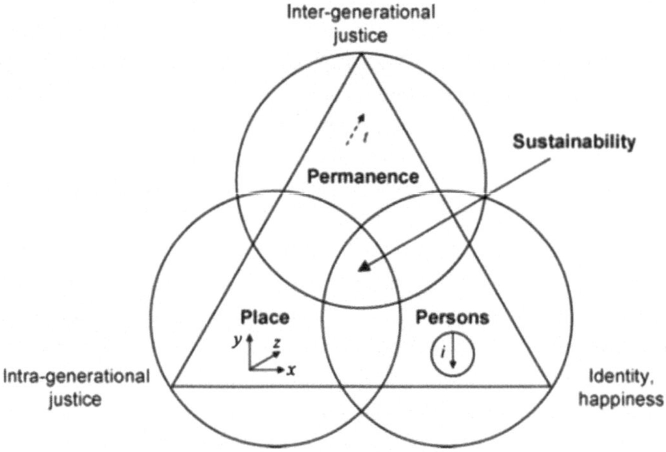

Fig. 4 The five-dimensional sustainability triangle (*Source* Seghezzo, 2009)

Urban Sustainability and Culture

As mentioned earlier, Seghezzo (2009) argues that sustainability has five dimensions; *place*, *permanence* and *persons*. The first three dimensions corresponds to the place, as explained earlier is the space where the identity of the place and the cultural values are transmitted. Researchers in the field of culture, such as Holahan and Bonnes Dobrowolny (1979), Kent (1990) and Rapoport (1998) and argued for the definition of a place by individuals' activities occuring in it, of which many may not be disturbed as time passes, thereby placing emphasis on theoretical links between person, place, and time through constantly repeated activities and leading to the second dimension, permanence which adds time to place in that place is at heart a cultural construct related to past—sometimes long past - events. Lawrence and Low (1990) and other design and behaviour studies advocates stress the importance of time's role which must be considered if time's spatial, cultural, social and personal Dimensions are to be understood. Finally, the persons dimension sees individuals as part of society instead of ignoring them when society is conceptualised. This awareness of the individual entity within its close relationship with the concept of place are aligned to Graumann's definition of appropriation already mentioned in the INTRODUCTION section. Thus, by adding the dimension of time, this suggests a strong link may exist between the concept of TA and urban sustainability.

Past Studies About Social Sustainability

Studies with focus on social sustainability have been accomplished at different urban scales. For instance, Forman (1997), Knippenberg et al. (2007), Wheeler (2000) and at regional level; Burton (2000), Dempsey et al. (2011), Dixon et al. (2009), Girardet (1999), Nijkamp and Perrels (1994), Roseland (1997), Sharifi and Murayama (2013) and Yiftachel and Hedgcock (1993) at the city level; Ancell and Thompson-Fawcett (2008), Bramley and Power (2009), Chiu (2003), Opp (2017), Pongsmas (2004) and Raman (2010) at neighbourhood level. However, authors (Mehan & Soflaei, 2017; Shirazi & Keivani, 2017) argue that studies focused on urban social sustainability at the microscale level (public space) are the less explored. For instance, studies focussed on squares exploring strategies for regeneration aiming to achieve social sustainability (Mehan, 2016) and studies focused on determining the social function of the square (Harun, Zakariya, Mansor, & Zakaria, 2014). Mehan (2016) concludes that the preservation of the historical identity is one of the major contributing approaches of urban regeneration strategies to achieve social sustainability in public squares with historical value. Results of a multi-layered study by Harun et al. (2014) suggest that a high diversity-range of activities in which people temporarily appropriate the public space contributes to the social sustainability of the site. Hence, the physical dimension of the environment and the TA of the public space are associated with social sustainability. Mehan and Soflaei (2017) attempted to link

Table 2 Translation of the human needs to spatial qualities in an urban context (*Source* Mehan & Soflaei, 2017)

Human needs residents	Spatial qualities in urban design
Physiological needs	Comfort/publicServices/firmness and balance
Safety needs	Privacy/legibility/safety
Belongingness/love needs	Social amenities/social facilities/sense of place and identity
Esteem needs	Inclusiveness/preservation of the local characteristics
Self-actualization needs	Diversity/public participation
Beauty needs	Visual richness/visual proportions/visual distinctiveness

theory and practice in the social sustainability field by incorporating Maslow's hierarchy of human needs into the urban environment. They argue that spatial qualities of the design of the built environment can be made in reflecting these needs. Table 2 summarises their claims.

Despite this attempt, Maslow's theory has a fundamental critique. First, it has been argued that Maslow's theory only applies to pluralistic societies (Western) which make it culture-specific rather than universal in the application (Neher, 1991). Secondly, in advanced societies, the first two steps of Maslow's needs are often satisfied, whereas the following steps (love and self-esteem) represent a challenge; contrary to what occurs in simpler societies (Turnbull, 1974). Therefore, it could be risky to apply such an approach in cities that hold a multi-diverse and ancient cultural background, such as Latin American, Mediterranean or Asian cities.

The Missing Link

The previous sections explain how TA and urban social sustainability are interrelated concepts, in which the cultural factor plays a determinant role. Figure 5 provides a summary and illustration of this relationship. It uses colour to differentiate between different fields of research into both concepts. These include architecture which itself embraces urban planning, landscape and urban design, as well as environmental and social sciences. It identifies source types based on relevance to this text including key paper, seminal paper and theory. Finally, it illustrates links between sources and their relevance, with the degree of relevance being signified by the darkness in colour of the name and symbol.

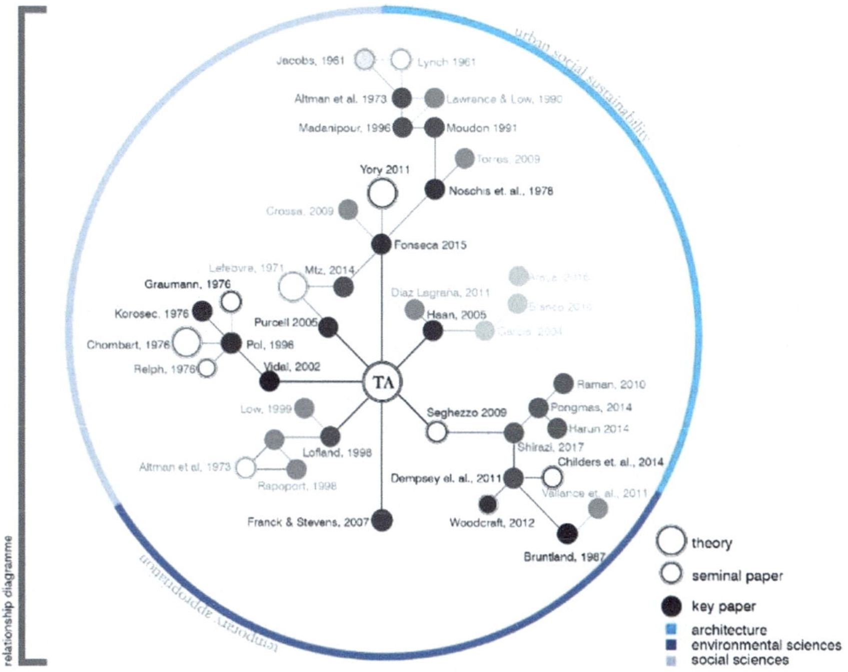

Fig. 5 Temporary appropriation and urban social sustainability diagram (*Source* Lara-Hernandez & Melis, 2018)

Discussion and Conclusion

An initial objective of this chapter was to illustrate the relationship between two complementary concepts, the TA and the social sustainability. The first section has discussed the relevance of the TA for the social construction of the urban landscape. It has been explained that the TA occurs in public spaces with meanings and values being socially constructed, which are places. In addition, it has categorised drivers into three main groups by illustrating examples from different cities and cultures. Culture and TA are related both dynamically and reciprocally, with time being crucial in forming, transitioning and preserving cultural values in the urban environment.

The second section established urban social sustainability as one of the key concepts for sustainable development. Social sustainability is multi-disciplinary and multidimensional and correlates strongly with design and other aspects of the built environment and these multidimensional aspects bring a complexity that permits research that is specific to the case and the place, suggesting social sustainability studies should relate to every particular context. Several authors (Antrop, 2005; Clark & Dickson, 2003; Kates et al., 2001) all believe that sustainability studies would be improved by multidisciplinary approaches such as this one, so that studies of the TA of public space are strongly related to the urban landscape's social aspect, and therefore specific to the context in question.. Moreover, the improvement and maintenance of the physical dimension of the environment are widely accepted as a strategy to achieve social sustainability.

However, further studies need to be carried out in this field of research.

Overall, this study strengthens the idea that the TA could be an indicator to assess the urban social sustainability in any specific context. These findings have significant implications for the understanding of how the design of the built environment could have the capability to allow the TA for future generations, which is one of the main aims of the urban sustainability. This work contributes to existing knowledge of social sustainability by providing the TA as a medium to study the urban landscape at the scale of the public spaces. Within the field of architecture (landscape, urban design and planning), more information on the design features of the built environment related to the TA in any determinate context would help us to establish a greater degree of accuracy on this matter. The scope of this study was limited in terms of urban scale due to it being focused at the level of the streetscape. In terms of direction for further research, further work could be focussed in other urban scales such as neighbourhood, districts and cities. It would be interesting to identify and categorise the TA in cities with different cultural characteristics and weather conditions than the ones illustrated here. It will be interesting to assess the effects of TA in relation to well-being, sense of community or even crime perception. There is, therefore, a definite need for a deep understanding of the TA and its relationship with the sustainability of any socio-cultural landscape.

References

Ali-Toudert, F., & Ji, L. (2017). Modeling and measuring urban sustainability in multi-criteria based systems—A challenging issue. *Ecological Indicators*, *73*, 597–611. [Online]. Available from https://doi.org/10.1016/j.ecolind.2016.09.046.

Ancell, S., & Thompson-Fawcett, M. (2008). The social sustainability of medium density housing: A conceptual model and Christchurch case study. *Housing Studies*, *23*(3), 423–442. [Online]. Available from https://doi.org/10.1080/02673030802029990.

Antrop, M. (2005). Why landscapes of the past are important for the future. *Landscape and Urban Planning*, *70*(70), 21–34. [Online]. Available from https://doi.org/10.1016/j.landurbplan.2003.10.002.

Barton, A. C. (2002). Urban science education studies: A commitment to equity, social justice and a sense of place. *Studies in Science Education*, *38*(1), 1–37. [Online]. Available from https://doi.org/10.1080/03057260208560186.

Bonfil Batalla, G. (2004). Pensar nuestra cultura. *Dialogos de accion, primera etapa*, 117–134.

Bramley, G., & Power, S. (2009). Urban form and social sustainability: The role of density and housing type. *Environment and Planning B: Planning and Design*, *36*(1), 30–48. [Online]. Available from https://doi.org/10.1068/b33129.

Brundtland, G. H. (1987). *World commission on environment and development (1987): Our common future*. Oxford University Press.

Burton, E. (2000). The potential of the compact city for promoting social equity. In Elizabeth Burton, Mike Jenks, & Katie Williams (Eds.), *Achieving sustainable urban form* (pp. 19–29). New York: Routledge.

Cambridge English Dictionary. (2018). *Culture meaning in the Cambridge English Dictionary*. [Online]. Available from https://dictionary.cambridge.org/dictionary/english/culture. Accessed January 22, 2018.
Capel Sáez, H. (2002). *La morfología de las ciudades*. Barcelona: España, Ediciones del Serbal.
Childers, D. L., Pickett, S. T. A., Grove, J. M., & Ogden, L. et al. (2014). Advancing urban sustainability theory and action: Challenges and opportunities. *Landscape and Urban Planning, 125*, 320–328. [Online]. Available from https://doi.org/10.1016/j.landurbplan.2014.01.022.
Chiu, R. L. H. (2003). Social sustainability, sustainable development and housing development: The experience of Hong Kong. In Ray Forrest & James Lee (Eds.), *Housing and social change: East-west perspectives* (p. 221). London: Routledge.
Clark, W. C., & Dickson, N. M. N. (2003). Sustainability science: The emerging research program. In *Proceedings of the National Academy of Sciences of the United States of America, 100*(14), 8059–8061. [Online]. Available from https://doi.org/10.1073/pnas.1231333100.
Costanza, R. (1992). *Ecological economics: The science and management of sustainability*. Columbia University Press.
Crossa, V. (2009). Resisting the entrepreneurial city: Street vendors' struggle in Mexico City's historic center. *International Journal of Urban and Regional Research, 33*(1), 43–63. [Online]. Available from https://doi.org/10.1111/j.1468-2427.2008.00823.x.
Crossa, V. (2016). Searching for order on the 'dirty' streets of the city. Displacement of street vendors in colonial and postcolonial Mexico City. In *Contested Cities*. Madrid, España. p.
de la Torre, M. I. (2015). Espacio público y colectivo social. *Nova Scientia, 14*(7), 495–510.
Deakin, M., Mitchell, G., Nijkamp, P., & Vreeker, R. (2007). *Sustainable urban development volume 2: The environmental assessment methods*. Routledge.
Dempsey, N. (2006). *The Influence of the quality of the built environment on social cohesion in English neighbourhoods*. Oxford Brookes University.
Dempsey, N., Bramley, G., Power, S., & Brown, C. (2011). The social dimension of sustainable development: Defining urban social sustainability. *Sustainable Development*, 289–300. [Online]. Available from https://doi.org/10.1002/sd.417. May 19, 2009.
Díaz Larrañaga, N., Grassi, L., & Mainini, C. (2011). Socialidad: Los modos de apropiación del espacio público. *Question, 1*(29), 12. [Online]. Available from http://perio.unlp.edu.ar/ojs/index.php/question/article/view/572/849.
Dixon, T., Colantonio, A., Ganser, R., & Carpenter, J. et al. (2009). Measuring socially sustainable urban regeneration in Europe. *Oxford Institute for Sustainable Development (OISD) and European Investment Bank (EIB)*.
Drummond, L. B. W. (2000). Street scenes: Practices of public and private space in Urban Vietnam. *Urban Studies, 37*(12), 2377–2391. [Online]. Available from https://doi.org/10.1080/00420980020002850.
Eliot, T. S. (1949). *Notes towards the definition of culture*. [Online]. Available from https://books.google.co.uk/books?id=QYR9FNyKMmwC.
Fonseca Rodriguez, J. M. (2015). La importancia y la apropiación de los espacios públicos en las ciudades. *Revista de Tecnologia y Sociedad., 4*(7), 1–11.
Forman, R. (1997). *Land Mosaics: The ecology of landscapes and regions*. Cambridge: Cambridge University Press.
Franck, K. A., & Stevens, Q. (2007). *Loose space: Diversity and possibility in urban life*. New York: Routledge.
Girardet, H. (1999). *Creating sustainable cities*. Resurgence Books.
Graumann, C. (1983). On multiple identities. *International Social Science Journal, 35*(96), 309–321. [Online]. Available from http://eds.b.ebscohost.com/eds/pdfviewer/pdfviewer?sid=4e7bdb42-733e-4fbb-98b0-3082b60477c5%40sessionmgr105&vid=7&hid=112.
Graumann, C. F. (1976). *The concept of appropriation (aneignung) and modes of appropriation of space*, *6*(2), 301–313. [Online]. Available from http://iaps.scix.net/cgi-bin/works/Show?iaps_00_1976_009.

Haan, H. De (2005). *Social and material appropriation of neighborhood space : Collective space and resistance in a Dutch Urban community* (pp. 1–122).

Harun, N. Z., Zakariya, K., Mansor, M., & Zakaria, K. (2014). Determining attributes of urban plaza for social sustainability. *Procedia—Social and Behavioral Sciences, 153*, 606–615. [Online]. Available from https://doi.org/10.1016/j.sbspro.2014.10.093.

Hawkes, J. (2001). *The fourth pillar of sustainability: Culture's essential role in public planning*. Common Ground.

Hernandez Mendo, A., & Morales Sanchez, V. (2008). Una revisión teórica: ocio, tiempo libre y animación sociocultural. *Efdeportes.com, 27*, 9–12. [Online]. Available from http://www.efdeportes.com/efd127/una-revision-teorica-ocio-tiempo-libre-y-animacion-sociocultural.htm.

Holahan, C. J., & Bonnes Dobrowolny, M. (1979). Cognitive and behavioral correlates of the spatial environment. An international Analysis. *Environment & Behavior, 10*(3), 317–333.

Hubbard, P., Kitchin, R., Bartley, B., & Fuller, D. (2002). *Thinking geographically*. London: Continuum.

Kates, R. W., Clark, W. C., Corell, R., Hall, J. M., et al. (2001). Sustainability Science. *Science's Compass, 292*, 641–642.

Kent, S. (1990). *Domestic architecture and the use of space*. Cambridge: Cambridge University Press.

Kim, A. M. (2013). *Sidewalk city*. [Online]. US: The University of Chicago Press. Available from https://doi.org/10.7208/chicago/9780226119366.001.0001.

Kleine, A. (2009). *Operationalisierung einer Nachhaltigkeitsstrategie: Ökologie*. Ökonomie und Soziales integrieren: Springer-Verlag.

Knippenberg, L., Beckers, T., Haarmann, W., & Hermans, F. et al. (2007) Developing tools for the assessment of sustainable development in the province of Brabant, the Netherlands. *Sustainability Indicators-Scientific Assessment*, 309–328.

Kyle, G. T., Jun, J., & Absher, J. D. (2014). Repositioning identity in conceptualizations of human–place bonding. *Environment and Behavior, 46*(8), 1018–1043. [Online]. Available from https://doi.org/10.1177/0013916513488783.

Lara-Hernandez, J. A., & Melis, A. (2018). Understanding the temporary appropriation in relationship to social sustainability. *Sustainable Cities and Society, 39C*, 1–14. [Online]. Available from https://doi.org/10.1016/j.scs.2018.03.004.

Lara-Hernandez, J. A., Melis, A., & Caputo, S. (2017). Understanding spatial configuration and temporary appropriation of the street in Latin American cities: The case of Mexico City Centre Hocine Boudagh, Antonella Versaci, Ferdinando Trapani, Marco Migliore et al. (Eds.), *Advances in Science, Technology & Innovation*, 153–173.

Lawrence, D. L., & Low, S. M. (1990). The built environment and spatial form. *Annual Review of Anthropology. 19*(1), 453–505. [Online]. Available from https://doi.org/10.1146/annurev.anthro.19.1.453.

Leary-Owhin, M. (2015). A fresh look at lefebvre's spatial triad and differential space: A central place in planning theory? In *2nd Planning Theory Conference University of the West of England* (pp. 1–8).

Lofland, L. H. (1998). *The public realm: Exploring the city's quintessential social territory*. New York: Transaction Publishers.

Loukaitou-Sideris, A., & Ehrenfeucht, R. R. (2009). *Sidewalks conflict and negotiation over public space*. USA: Massachusetts Institute of Technology.

Lozano, R. (2008). Envisioning sustainability three-dimensionally. *Journal of Cleaner Production, 16*(17), 1838–1846. [Online]. Available from https://doi.org/10.1016/j.jclepro.2008.02.008.

MacDonald, R., & Shildrick, T. (2007). Street corner society: Leisure careers, youth (sub)culture and social exclusion. *Leisure Studies, 26*(3), 339–355. [Online]. Available from https://doi.org/10.1080/02614360600834826.

Madanipour, A. (1996). *Design of urban space. An inquiry into socio-spatial process*. New York: John Wiley & Sons Inc.

Martínez, E. (2014). Configuración urbana, hábitat y apropiación del espacio. *Scripta Nova. Revista electrónica de geografía y ciencias sociales, XVIII*(493), 20.
Maslow, A.H. (1943). A theory of human motivation. *Psychological review, 50*(4), 370.
Mehan, A. (2016). Urban regeneration: A comprehensive strategy for achieving social sustainability in historical squares. In *3rd International Multidisciplinary Scientific Conference on Social Sciences and Arts. 2016 SGEM International Multidisciplinary Scientific Conference on Social Sciences And Arts* (pp. 861–868). [Online]. Available from https://doi.org/10.5593/sgemsocial2016hb42.
Mehan, A., & Soflaei, F. (2017). Social sustainability in urban context. In *Architectural Research Addressing Societal Challenges* (pp. 293–299). [Online]. CRC Press. Available from doi: https://doi.org/10.1201/9781315226255-47.
Monnet, J. (1995). *Usos e imágenes del centro histórico de la Ciudad de México*. Centro de estudios mexicanos y centroamericanos (Ed.).
Moudon, A. V. (1991). *Public streets for public use*. New York: Columbia University Press.
Mouffe, C. (2014). *Agonística. Pensar el mundo políticamente*. Soledad Laclau (Ed.), Mexico, DF, Fondo de Cultura Económica de Argentina.
Munasinghe, M. (1993). *Environmental economics and sustainable development*. World Bank Publications.
Musacchio, L. R. (2009). The scientific basis for the design of landscape sustainability: A conceptual framework for translational landscape research and practice of designed landscapes and the six Es of landscape sustainability. *Landscape Ecology, 24*(8), 993–1013. [Online]. Available from https://doi.org/10.1007/s10980-009-9396-y.
Neher, A. (1991). Maslow's theory of motivation: A critique. *Journal of Humanistic Psychology, 31*(3), 89–112. [Online]. Available from https://doi.org/10.1177/0022167891313010.
Nijkamp, P., & Perrels, A. (1994). *Sustainable cities in Europe*. London: Earthscan.
Opp, S. M. (2017). The forgotten pillar: A definition for the measurement of social sustainability in American cities. *Local Environment, 22*(3), 286–305. [Online]. Available from https://doi.org/10.1080/13549839.2016.1195800.
Oxford University. (2018). *Culture/Definition of culture in English by oxford dictionaries*. [Online]. Available from https://en.oxforddictionaries.com/definition/culture. Accessed January 22, 2018.
Pol, E. (2002). The theoretical background of the city-identity-sustainability network. *Environment & Behavior, 34*(1), 8–25.
Pol, E., & Iñiguez, L. (1996). La Apropiacion del Espacio. *Monografías Psicosoaicles Ambientales, 9*.
Pol Urrútia, E. (2002). El modelo dual de la apropiación del espacio en Mira, R., Sabucedo, J. M y Romay, J. *Psicología y Medio Ambiente. Aspectos psicosociales, educativos y metodológicos* (pp. 123–132).
Pongsmas, N. (2004). *Configuration of public space and social sustainability of urban neighborhood: A case study of the city of San Diego at the dawn of the twenty-first century*. [Online]. Texas Tech University. Available from https://ttu-ir.tdl.org/ttu-ir/handle/2346/19502.
Portal, M. A. (2009). Las creencias en el asfalto. La sacralización como una forma de apropiación del espacio público en la ciudad de México. *Cuadernos de Antropología Social, 30,* 59–75.
Raman, S. (2010). Designing a liveable compact city physical forms of city and social life in urban neighbourhoods. *Built Environment, 36*(1), 63–80. [Online]. Available from https://doi.org/10.2148/benv.36.1.63.
Rapoport, A. (1998). Using 'Culture' in housing design. In *Housing and society*. Korea (pp. 1–20).
Rapoport, A. (2005). *Culture, architecture, and design*. Chicago: Locke Science Publisher.
Relph, E. (1976). *Place and placelessness* (2nd edn.). Pion Ltd.
Roseland, M. (1997). Dimensions of the eco-city. *Cities, 14*(4), 197–202. [Online]. Available from https://doi.org/10.1016/s0264-2751(97)00003-6.
Seaman, J., & McLaughlin, S. (2014). The importance of outdoor activity and place attachment to adolescent development in Coös County, New Hampshire. *Building knowledge for families and communities, 208*(37). [Online]. Available from http://scholars.unh.edu/carsey/208/.

Seghezzo, L. (2009). The five dimensions of sustainability. *Environmental Politics*, 18(4), 539–556. [Online]. Available from https://doi.org/10.1080/09644010903063669.

Sharifi, A., & Murayama, A. (2013). A critical review of seven selected neighborhood sustainability assessment tools. *Environmental Impact Assessment Review*, *38*, 73–87. [Online]. Available from https://doi.org/10.1016/j.eiar.2012.06.006.

Shirazi, M. R., & Keivani, R. (2017). Critical reflections on the theory and practice of social sustainability in the built environment—a meta-analysis. *Local Environment*, 1–20. [Online]. Available from https://doi.org/10.1080/13549839.2017.1379476.

Simpson, P. (2011). Street performance and the City: Public space, sociality, and intervening in the everyday. *Space and Culture*, *14*(4), 415–430. [Online]. Available from https://doi.org/10.1177/1206331211412270.

Spindler, E. (2011). Geschichte der Nachhaltigkeit Vom Werden und Wirken eines beliebten Begriffes. *Edmund A. Spindler*, *21*.

Torres, E. (2009). *Apropiación Versus Usos Del Espacio Público*. [Online]. Available from https://www.mardelplata.gob.ar/documentos/policia/taller1culturayterritorioapuntes.pdf. Accessed November 1, 2017.

Turnbull, C. M. (1974). *The mountain people*.

Valentin, A., & Spangenberg, J. H. (2000). A guide to community sustainability indicators. *Environmental Impact Assessment Review*, *20*(3), 381–392. [Online]. Available from https://doi.org/10.1016/s0195-9255(00)00049-4.

Vallance, S., Perkins, H. C., & Dixon, J. E. (2011). What is social sustainability? A clarification of concepts. *Geoforum*, *42*(3), 342–348. [Online]. Available from https://doi.org/10.1016/j.geoforum.2011.01.002.

Vidal Moranta, T., & Pol Urrútia, E. (2005). La apropiación del espacio: Una propuesta teórica para comprender la vinculación entre las personas y los lugares. *Anuario de Psicologia*, *36*(3), 281–297. [Online]. Available from https://doi.org/10.1006/jevp.2000.0185.

Vojnovic, I. (2014). Urban sustainability: Research, politics, policy and practice. *Cities*, *41*, S30–S44. [Online]. Available from https://doi.org/10.1016/j.cities.2014.06.002.

Wheeler, S. M. (2000). Planning for metropolitan sustainability. *Journal of Planning Education and Research*, *20*(2), 133–145. [Online]. Available from https://doi.org/10.1177/0739456x0002000201.

Woodcraft, S. (2012). Social sustainability and new communities: Moving from concept to practice in the UK. *Procedia—Social and Behavioral Sciences*, *68*, 29–42. [Online]. Available from https://doi.org/10.1016/j.sbspro.2012.12.204.

Woodcraft, S. (2015). Understanding and measuring social sustainability. *Journal of Urban Regeneration & Renewal*, *2*, 133–144. [Online]. Available from http://www.ingentaconnect.com/content/hsp/jurr/2015/00000008/00000002/art00004.

Yiftachel, O., & Hedgcock, D. (1993). Urban social sustainability. *Cities*, *10*(2), 139–157. [Online]. Available from https://doi.org/10.1016/0264-2751(93)90045-k.

Between Assemblages and Temporary Appropriation: The Case of Mexico City

Jose Antonio Lara-Hernandez, Alessandro Melis and Steffen Lehmann

Abstract The chapter describes temporary appropriation (TA) as an assemblage product of other assemblages within the urban landscape. It unravels and invites to re-think the nature of TA through assemblage theory. Derived from the work of Deleuze and Guattari (A thousand plateaus: Capitalism and schizophrenia. University of Minnesota Press, Minneapolis, USA, 1989) and developed further by Manuel DeLanda (Assemblage theory. Edinburgh University, Edinburgh, UK, 2016), assemblage theory focuses on the relations produced by the components of a whole rather than the components themselves. Thus, the present chapter combines a range of theories to conceptualise TA of the urban landscape as an emerging product of other assemblages such as culture, legal framework and urban design. These approaches are drawn together by illustrating Mexico City Centre as an example of a highly coded city in which these assemblages emerge. A representative sample street was selected as a case-study to analyse TA in relation to the streetscape design through participant observation and image analysis of the visual complexity of the streetscape. The chapter concludes that assemblage theory could be used as a theoretical framework investigating urban-social phenomena. In addition, the study identified the visual complexity of the assemblage of the urban landscape that supports the greater diversity of TA.

Keywords Assemblage theory · Social sustainability · Temporary appropriation · Urbanism

Introduction

In total, 66% of the urban population today live in slums or informal settlements. This proportion is expected to increase to 70% by 2050, whilst the total urban population will include 80% of the world population (United Nations, 2014). In such a scenario

J. A. Lara-Hernandez (✉) · A. Melis
University of Portsmouth, Portsmouth, UK
e-mail: Jose.LaraHernandez@myport.ac.uk

S. Lehmann
University of Nevada, Las Vegas, Las Vegas, USA

A. Melis et al. (eds.), *Temporary Appropriation in Cities*,
https://doi.org/10.1007/978-3-030-32120-8_3

the informal use of public spaces in a compact city, argued to be a more sustainable urban model (Bay & Lehmann, 2017), will play a decisive role. TA is a key concept related to the informal use of public space. As stated in previous chapter, TA of the urban landscape plays a decisive role in sustainability in its social dimension (Lara-Hernandez & Melis, 2018; Seghezzo, 2009; Ramirez Kuri, 2010; Marx & Kelling, 2018; Devlin, 2017). Consistent with this premise, its relevance will grow in conjunction with the dramatic growth of the informal part of the urban population. The street's multidimensional use will be favoured by risks that urban open spaces in a compact city will become scarce (Ekawati, 2015). Be that as it may, there is a shortage of knowledge about what a streetscape's TA really is and how it contributes to the urban landscape's social sustainability.

This chapter serves for two main purposes, first describes the relationship between the urban design landscape and behavioural patterns, through the study of TA. Considering the novelty of the subject and its interdisciplinary nature, to address this research aim the chapter uses an assemblage thinking approach which focuses on the relations between the assemblage's components rather than on the individual components themselves. In this specific case, it identifies three main components which are also assemblages (to be discussed further). Firstly, TA of the streetscape as an emergence indicator of the informal use of the urban landscape; secondly, culture as a factor in social sustainability and lastly, the legal framework as a component (and an assemblage) regulating the fruition of public space. Second, it shows TA as part of the urban landscape and as an emerging assemblage product of other assemblages or "social wholes", such as cultural and legal frameworks. Therefore the focus of the present chapter is the relationships and interconnections between TA and assemblage theory in the field of urban sciences. As already explained, TA is a re-emerging concept which occurs in the urban social landscape as a multidimensional phenomenon. Intended as multi-disciplinary and multi-scalar research, the present chapter explores the way in which TA could be interpreted as (but not limited to) an emergence assemblage product of other assemblages such as the cultural, the legal and urban design landscape of what DeLanda (2016) describes as a highly coded city. To do so, a street in Mexico City Centre (MCC) was selected as a sample to illustrate this in detail. Lastly, the chapter argues that TA could play an important role towards environmental sustainability by building towards energy reduction and public health by increasing outdoor activities. The following sections explain the assemblage thinking approach as a conceptual framework.

Assemblage Thinking and Territorialisation

The use of assemblage theory to explore, analyse or describe urban phenomena has become significant today thanks to its orientation towards considering different disciplines. The literature ranges from more theoretical pieces such as Anderson, Kearnes, McFarlane, and Swanton (2012), Müller and Schurr (2016) and Dovey et al. (2018) to those more focused on its practical application such as McFarlane

(2011), Dovey (2012) and Dovey and Pafka (2014). Deleuze and Parnet define an assemblage as a "multiplicity which is made up of many heterogeneous terms and which establishes liaisons, relations between them, across ages, sexes and reigns—different natures. Thus, the assemblage solely unit is that of a co-functioning: it is a symbiosis, a sympathy" (Deleuze & Parnet, 1977: p. 52). Assemblage theory uses a flat ontology, which is a model for reality that states that all object (even imaginary), have the same degree of being-ness as any other object. There is no hierarchy among the objects. All subjects are simply objects. The key factor in determining ontology is the ability of an object to affect another object (Kramer, 2019). Assemblage thinking is less concerned with the elements assembled than with the relationships between them and seeks an understanding of how combined flows and synergism collaborate. In addition, DeLanda (2016) modifies Deleuze and Guattari's original concept, regarding as assemblages parts forming an ensemble and having their own parameters so that we are always dealing with assemblages of assemblages. This approach looks for connections between categories or thought clusters that are different. An assemblage is merely an arrangement and does not have the hierarchy expected in a system. Entities are interconnected but no hierarchy is apparent. The definition offered by Marston et al. (2005) is "flat ontology". This is the opposite of moving from the particular to the general (Dovey, Rao, & Pafka, 2018), with whom we agree that scholars usually use urban environment research to make a theoretical point, while assemblage thinking's main concern should be the use of theory as the way to understand and improve the city.

Within urban research studies, the understanding of "territorialisation", a key term in assemblage theory, is essential. Territorialisation defines the degree of homogenisation of assemblage components and how far boundaries have been shown and made impenetrable (DeLanda, 2016). On the one hand, this is a process of defining and sharpening actual territories' spatial boundaries. On the other, it describes non-spatial processes that increase an assemblage's homogeneity. The term deterritorialisation is applied to processes that either increase internal heterogeneity or destabilise spatial boundaries. Territorialisation is regarded in assemblage theory as the components' first articulation while genes, words or architectural elements encode the second articulation which stabilises the assemblage's identity. From assemblies of smaller entities, larger entities emerge. Assemblage theory has two main features (Müller & Schurr, 2016). The first is that the parts making up assemblages are self-sufficient and articulated by exterior relationships of exteriority and can therefore be detached and become a component in another assemblage. The second is that assemblages have two dimensions of which the first is the parts' variable roles and the second is the convoluted nature of the processes in which components are engaged with may stabilise or destabilise the assemblage. A third dimension was added by DeLanda (2011) in the shape of an axis by which processes are defined that consolidate the assemblage's identity or which puts it into a flexible operational state. It is by the addition of this axis that the theory of how assemblages are products of other assemblages can be explained.

Another important component of assemblages is "social wholes". According to DeLanda (DeLanda, 2006, 2016), a social whole is an interpersonal network or instructional organisation which can neither be reduced nor totalised and is a community of which the emergent property is the extent of the between its members who may be persons or institutions. McFarlane (2011) describes the common conceit of assemblage by emphasising how elements join together and assemblages are understood as an orientation towards an object that operates as a way of studying the development of social, political and economic processes. In conclusion, we can think of assemblage as an orientation of the world and as an object in the world, since they are not mutually exclusive. What begins an assemblage is synergy between its components. Once in place, it acts both to limit and to facilitate components, in a process called by DeLanda (2016: p. 21) "downward causality". It could be argued that TA as an assemblage emerges from its cultural context, physical environment and the regulatory or legal framework that allows its occupation of part of the urban landscape (see 'The appropriation of the built-environment'). Spaces occupied in this way constrain some of their human components at the same time as they assist in their conformation as a cultural entity in a urban landscape bounded by desire which, thought of in this way, acts to attract parts of an assemblage to each other so that multiple desires can intersect and even contradict each other. Dovey et al. (2018) give an example by claiming that "the suburb emerges from desires to have day-to-day access to the city without living in it". Thus, we could argue that the desire to leave our trace in this world or simple spatial being produces TA of the urban landscape in any given context.

According to Anderson et al. (2012), assemblage theory is a way of conceptualising entities' interrelationships and the elements that make them up. The term allows the classification and categorisation of such amalgamated phenomena as TA. It has been used in other research fields such as ecology, art and archaeology. According to Anderson et al. (2012), an assemblage is an "experimental condition" for a theory of the arrangement of spatial and social aspects implying that analysis of composition uses assemblage as a concept to trigger an ethos of open engagement with the world. Examples would include the unity's form, the types of relationships involved and the interactions between the components. For instance, in terms of planning, the perimeter block in which the assemblage between planning and design, planning policy and inhabitants have shown to be far superior regarding social interactions among residents than suburban areas (Fleming, 1986; Raman, 2010). The relations between these three components are shown to be better environments in terms of mental health and wellbeing.

Jessop, Brenner and Jones (2008) see four dimensions to the spatial and social relations making up an assemblage: territory; scale; place; and network. For each dimension, the level of articulation and implication can be analysed in a specific "spatiotemporal context". Anderson et al. (2012), on the other hand, found fault with this idea. In the first place, its reduction of the relationships to patterns that are known and easy to identify prefigured a number of patterns principles and forms that was finite. In the second, assemblage allows formation with no requirement for a pre-existing configuration of relationships. Under the theory of assemblage, there can be different points of view about what and where a place is that can assist analysis of an urban landscape's identity, and particularly the built environment's influence on TA's emergence. For our part, we concur with Garcia (2013: p. 45) who said that the theory of assemblage did not explicitly state how processes depicting multiplicities operated, how an assemblage's elements interacted or how an urban landscape's identity responds to changes in the built environment. Thus assemblage theory could be enriched by other theoretical approaches in regard to social-urban research topics, such as the one presented here. In summary, it has been shown from this review that assemblage thinking is focused on the relations between components from which it emerges. An assemblage thinking approach means that even though one or more components of the assemblage could be changed or removed it does not necessarily affect the relations of the whole. Having defined what is meant by assemblage theory we will now move on to discuss TA.

Temporary Appropriation and Public Space

It has been explained previously that there is no agreement about the meaning of the term appropriation and uncertainty remains within social-urban design research fields. For instance, authors such as Garcia Ramon, Ortiz, and Prats (2004), Ramirez-Lovering (2008), Blanco, Bosoer, and Apaolaza (2014), Araya Diaz (2016) and Marx and Kelling (2018) refer to the term to informal or illegal use of a place or territory. However, we argue that TA should not be described as an informal or illegal activity in a public space, since the benefit of its citizens is the purpose of public space (further dicussed in Chapter "Temporary Appropriation and Informality"). The appropriation of the built environment is an socio-spatial demand innate to the individual (Graumann, 1976; Yory, 2011). Lefebvre (1971) supported this argument with three ways in which urban space is produced, and Yory (2011) added his support by saying that a commitment to construct and appropriate our surroundings is inborn in humans. Purcell (2002), Yory García (2003), Hernández-Bonilla and Gomez-Gomez (2015), Portal (2009) and Haan (2005) refer to this process as people's temporary activities in urban public spaces that help to construct urbanity, social cohesion, sense of belonging and identity for a specific place.

TA as a concept helps us to differentiate between activities in which individuals appropriate space from those activities that are simply the use of space (Lara-Hernandez & Melis, 2018). Figures 1 and 2 illustrate how activities related to the economy, which are extended from the place of working into the public realm (street), are an example of TA (Crossa, 2009, 2016; de la Torre, 2015; Kim, 2013; Lara-Hernandez, Melis, & Caputo, 2017; Loukaitou-Sideris & Ehrenfeucht, 2009; Monnet, 1995; Moudon, 1991; Vázquez & Tapia Quevedo, 2011). Figures 3 and 4 illustrate people carrying out leisure activities, which are certainly a type of activity related to TA (Drummond, 2000; Díaz Larrañaga, Grassi & Mainini, 2011; Hernandez Mendo & Morales Sanchez, 2008; Lara-Hernandez, Melis, & Caputo, 2017; MacDonald & Shildrick, 2007; Mouffe, 2014; Seaman & McLaughlin, 2014; Simpson, 2011; Lehmann, 2009; Groth & Corijn, 2005; Mouffe, 2012). Moreover, TA is also manifested through religious activities (Fig. 5). At first glance, these activities might

Fig. 1 (left) Man selling chestnuts in the street in the centre of Palermo; (right) Man selling handcrafted jewellery in the centre of Mexico City (*Source* Authors)

Fig. 2 (left) Woman selling soup in Ho Chi Minh City, Vietnam; (right) Man and boy selling water, Ho Chi Minh City, Vietnam (*photos* Antonino DiRaimo)

Fig. 3 Kids skateboarding on a street corner in Chichester, UK

Fig. 4 (left) Men playing cards in the street in the historic districts of Palermo; (right) Man watering his "garden" the street in the centre of Mexico City (*Source* Authors)

Fig. 5 (left) Couple preaching in the street in Auckland CBD; Altar in the street of the centre of Mexico City (right) (*Source* Authors)

be seen to be associated with underdevelopment or even informality, but in terms of social-urban dynamics they are more than that. The following paragraphs will explain this in more detail.

As a socio-spatial phenomenon forming part of a city's cultural landscape, TA is multi-dimensional and dynamic. Multi-dimensional because, far from simply occupying a space, it takes place in what, in the words of Relph (1976), combines natural and human dimensions and how we experience the environment—it is at the centre of action. It is space full of meaning and endowed with symbols and significance; it is a social construct with a set temporal duration (Madanipour, 1996). It contains hints of the hidden meaning, rituals and symbols with which people imbue particular places (Rapoport, 2005). Dynamic because of the conditioned temporality implicit in the way Yory's interprets the theory of topophilia already explained. It is socio-spatial because, in any given context, TA results in the interaction between an individual and the social whole and its environment in the historical context.

As noted by Lara-Hernandez, Melis, and Caputo (2017) TA is an interaction between the individuals (or collectively) and their city through specific activities happening in the urban landscape. Moreover, this interaction embodies the cultural landscape in which people perform an important role in its configuration (Hubbard, Kitchin, Bartley, & Fuller, 2002; Seghezzo, 2009). Following this argument Graumann (1976, 1983), Pol Urrútia (2002), Vidal Moranta (2005) and recently, Martínez (2014) state that only public spaces in where people feel identified are temporarily appropriated.

Different authors have measured the appropriation of public spaces in the urban environment in a variety of ways. Whyte (1980), in his seminal work, analysed people's activities in public and semi-public squares in New York through the use of video cameras. Along the same argument but in the Latin-American context, Salazar-Trujillo (2010) and his team analysed the permanence, use and occupancy of squares and streets in Bogota. However, there are certain drawbacks associated with the use of video cameras in public spaces. First, when users notice them they tend to behave differently from usual (Salazar Trujillo, 2010) and secondly, they require more time, labour and technological resources.

Highly Coded City

Considering the cultural implications previously highlighted, the selection of the city is pivotal to better understand how TA could be explained as an assemblage of assemblages within the built-environment. According to DeLanda (2016) it is possible to describe some cities as coded, implying a clear identity and a unique regional culture. A global city has multiple geographies (Fyfe & Kenny, 2005) so that the relationship between the cultural, social and physical landscapes is highly dynamic (Lima, 2001). As cultural, legal and social aspects of city landscapes or assemblages overlap, they define one particular place (Hubbard et al., 2002). In support of this argument, Brenner (2005) claims that a "vertical" differentiation exists where social relations are embedded in the hierarchical scaffolding of territorial units arranged like Russian dolls where the global contains the supra-national which contains the national and so on down to the urban, the local, and the body. Thus, we have selected MCC as an example of a highly coded city, which is convenient for illustrating one of the arguments presented in this chapter.

Mexico City Centre is an emblematic palimpsest dense with cultural recall. It was built on the top of Tenochtitlan (Fig. 6, left), the capital city of the Aztec empire, whose urban pattern was compounded by blocks, streets and channels. The Aztecs used the channels for communication purposes, while the streets were reserved for a diversity of activities such as trade, leisure, religious celebrations or even sacrifices (Leon Portilla, 1995). According to Kent (1990), the Aztecs were a civilisation with high layers of social complexity reflected in the design and use of their cities.

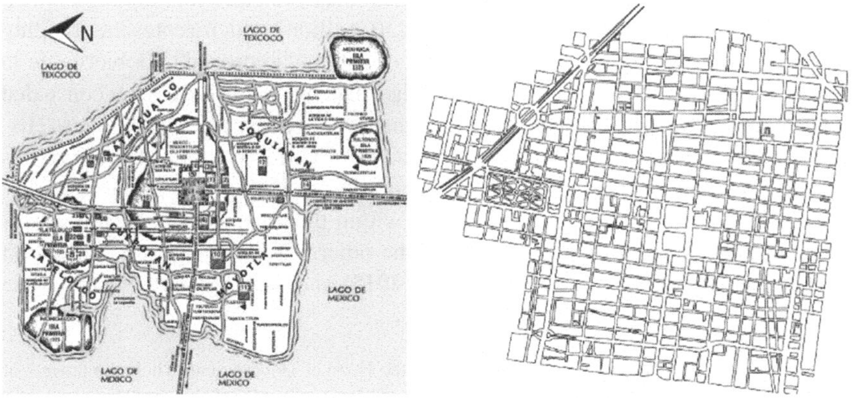

Fig. 6 (left) Tenochtitlan. Schematic representation 1325–15 (Carrera Stampa, 2002); (right) Mexico City Centre (Autoridad del Centro Historico, 2011)

The public spaces in which the social, political, economic and religious lives of people occurred were the streets (Webster & Sanders, 2001). Informal street life were particularly essential for the common population, confirming that pre-Hispanic civilisations had been characterised by intensively multiple use of outdoor spaces for everyday activities (Keller, 2006; Suárez Pareyón, 2004). In the sixteenth century, during the Spanish colonisation, a new order was established over Tenochtitlan, completely transforming the city in terms of its economic, social, cultural and legal dimensions (Stanislawski, 1947). Hence, the elements of the city or the assemblage were changed through a process of syncretism between the Spanish and Aztec cultures. Regarding TA, the conquerors tried to regulate informal activities (trading, playing, religious expressions) that happened on each specific street for each activity with a singular order (Nelson, 1963), by confining them to specific places, like squares in the Spanish tradition. They succeeded for a short period of time, but as the city grew, the confinement of informal activities was not viable anymore. As Monnet (1995, 1996) describes, this informality is still palpable in the streets of MCC. Currently, it is estimated that in Mexico City more than 1.2 million people are working in the informal economy (Gomez Flores, 2013). Culture structures behaviour and the use or non-use of the street. The streets of MCC are public spaces that are key to urban life and have been even prior to the Spanish colonisation.

MCC was the first colonial historic centre in America founded in 1521 which is also the biggest with 9.1 km^2. MCC's planning pattern was applied to most of the Spanish colonies in America (Stanislawski, 1947). In 1987 MCC was declared by UNESCO (2017) as a world heritage place. After the nomination, the government in association with private institutions have transformed and "improved" the built environment in MCC through urban design interventions. The built environment is one of the elements that contributes to the improvement of the quality of urban life (UNESCO, 2013). These interventions have as a main purpose to improve the quality of the built environment and therefore, as widely accepted, to increase urban-life standards such as community sense, inclusion, pluralistic function and public space democracy. Between 2007 and 2014 there has been an economic investment of approx. 6 million Mx pesos (approx. £230 million) which represents roughly 82,579 m^2 of public spaces and 10.3 L/km of streets "improved" (Gobierno de la Ciudad de Mexico, 2016). These urban design improvements have been concealed and agreed between government, academics and residents (Autoridad del Centro Historico, 2011). On the one hand the Autoridad del Centro Historico[1] (2011) argues that there have been positive results such the improvement of the physical quality of the built environment, economic growth within the area, and the wealth of residents and formal workers in the area. On the other scholars such as Nivon-Bolan and Sanchez-Bonilla (2014), Flores-Arias (2015) and Ramirez-Kuri (2015) argue

[1] This is an institution that supports the activities of the Head of Government, where the powers of the Public Administration Units of the Government of Mexico City (CDMX) are concentrated by each delegation. It acts under the criteria of unity, autonomy, functionality, efficiency, coordination and impartiality for the planning and ordering of the territorial, economic and social development of the Historic Centre.

that such urban improvements in public spaces are leading to spatial exclusion and segregation. Thus, there are contradictory outcomes.

This section has attempted to provide a brief summary of the literature illustrating MCC as a highly coded city. We will now move on to explain two assemblages present in the urban landscape highly relevant to the emergence of TA as an assemblage.

Cultural Assemblage

TA is therefore triggered by the sense of cultural identification with a certain urban context. Tylor (1871) defines culture as a "complex whole which includes knowledge, beliefs, arts, morals, law, customs, and any other capabilities and habits acquired by [a human] as a member of society". In other words everything man-made could be described as part of culture. As a consequence of culture being a vast domain, this chapter takes into account only the built-environment as a cultural subset that is relevant to TA.

According to Guattari and Rolnik (2005), "culture" has had three different and still current usages throughout history. The first and the oldest describes how the spirit is cultivated. It signifies value and divides people between those who have culture and those who do not. The second is the culture-spirit and may refer, for example, to black culture, underground culture and technical culture. The third is culture-merchandise and describes the capital that produces cultural goods and may be money, people or institutions. The definition of culture given by Bonfil Batalla (2004) approached it more widely as a group of values, symbols, skills, attitudes, meanings, social structure, communication methods and physical objects that allow a determinate society a way of living. In addition, the latter allows the transformation and reproduction of culture through future generations. Moudon (1991) argues that the built-environment is a high-pitched display of cultural specificity by alleging that a distinct behaviour emerges from each type of society and, in terms of activities, is a reflection of it. For instance, in the Netherlands, public space is the space of strangers, since the borderline between the public and the private dimension of people's lives are very rigid (Haan, 2005), while in Mexico these boundaries can be blurred or overlap because public space is continually temporarily appropriated (Monnet, 1995).

Rapoport (1998) says that activities that take place in the public space, including TA, can be broken down into four components. These are: the activity itself; how it is conducted; its association with other activities and how it is integrated into activity systems; and what meaning can be deduced from the activity where the central factor is the space's design. The inference from Kyle, Jun, and Absher (2014) is that constant repetition of activities over time constructs urban identity and the bonding between individuals and place, an argument supported by Eliot (1949) who holds that these activities produce culture when, though each pursued for its own sake, they appear more or less harmoniously. DeLanda (2016) agrees with the latter by claiming that the habitual repetition of an action could be assumed to produce similar results in the future.

According to DeLanda (2006), not all social actions involve semantic interpretation. In some, tradition means that social activities associated may be close to the boundary of consciously oriented action, and may indeed be outside it. Moreover, "the practical routines could be somehow impregnated by ritualism symbolism, but at the same time being capable of leading to successful casual interactions with material entities" (DeLanda, 2016: p. 83). In this regard, TA is an assemblage in the same way that language is an assemblage because it is part of a and a parameter of a communal or organisational assemblage so that its existence is in the relationships between entities and not in the entities themselves (Hawkes, 1977). We are therefore free to regard culture as an assemblage and part of a dynamic continuum where culture is the result of activities repeated over time, though we must bear in mind that the same actions can also be a reflection of different cultures. While the elements of cultural assemblage have changed, it is the assemblage that is TA that has prevailed because the assemblages have retained their relationships.

Having defined TA and its dynamic reciprocal condition with culture, we will now move on to explain the legal framework as an assemblage of the urban landscape.

Legal Framework Assemblage

This section attempts to describe how the capacity of laws and regulations could act to affect the emergence of TA in MCC. The legal framework of the use of public spaces generates events that create social obligations, as incorporeal transformations that take place. According to Deleuze and Guattari (Deleuze & Guattari, 1989), a language that determines a body's capacity, most commonly as dictated by an authority, is legitimised institutionally or individually by the codes or regulations which constitute an important component playing along with as the buildings and human bodies that are the material components of the assemblage (DeLanda, 2016). For example, transforming a TA into an informal and illegal activity can be an instantaneous act or an incorporeal attribute that finds expression in the regulations.

Overall, the government of Mexico considers public space as an ambience or scenery for social integration, where there is a right of association and the right of others to use the same space, its appropriation (accessibility, permanence and enjoyment) of the space, the collective space, the space for everybody (Gobierno del Distrito Federal, 2013). However, TA is usually associated with informality and therefore, it constantly faces attempts to regulate it or forbid it, for instance, the Programa de Rescate (rescue program) of MCC in which one of the strategies was the removal of street vendors from perimeter A (Autoridad del Centro Historico, 2011). However, Crossa (2009) argues that in spite of the strategies implemented by MCC to remove informal commerce from the streets, street vendors have found ways to resist, and they have become *toreros* (this term refers to Mexico City's nomadic vendors), and are still working in the area.

Regulations such as *Ley para uso de las vías y los espacios públicos* (Gobierno del Distrito Federal, 2013), *Ley de la cultura cívica* (Asamblea Legislativa del Distrito

Federal III Legislatura, 2004), *Ley de Movilidad de la Ciudad de México* (Gobierno del Distrito Federal, 2016) and *Ley de Transito del Distrito Federal* (Gobierno del Distrito Federal, 2015) in Mexico City refer to the use of streets and the activities that are allowed to happen in them. All of them are components of the assemblages (already explained) which, stabilise or destabilise the assemblage through the process of territorialisation.

For instance, in 2004, the *Ley de la Cultura Civica del Distrito Federal* (Law of the Civic Culture of Mexico City) was approved by the government, and it essentially establishes how citizens should behave in public spaces. *Distrito Federal* (Federal District) was politically dependent upon the governor of the Estate of Mexico; however, after 2016 it is now part of Mexico City (Patiño, 2016). The 15th article of Chapter I/Second Title states that it will guarantee the harmony and coexistence of its inhabitants through the fulfilment of their duties, such as (a) the freedom of people's actions in public spaces, and (b) by allowing the proper use of public spaces according to their nature. Then in 2013, a more specific law was approved: *Ley para el uso de las vías y los espacios públicos del Distrito Federal* (Gobierno del Distrito Federal, 2013) (Law for the use of streets and public spaces in Mexico City), which principally establishes the right to use and enjoy public spaces, especially streets that are used in different ways, rather than just for motor-vehicles. On the one hand, Article 6 states that public space users (including users of the streets) have the right to use the property of common use, according to its nature, the right to access, stay in and transit the streets. On the other hand, Article 7 establishes that public space users have a duty to access, use, stay in or transit public space without disturbing other users. Thus, both laws make reference to and recognise the social dimension of the streets as public spaces in which TA emerges. Other laws, such as *Reglamento de Transito del Distrito Federal* (Transit Regulations of Federal District) and *Ley de Movilidad de la Ciudad de Mexico* (Law of Mobility of Mexico City) establish free access to streets and roads and the use of and transit of pedestrians, cyclists and motor-vehicles in the streets and roads.

The previous section has shown that the components of the assemblage (legal framework) in MCC have changed since colonisation and even in current times, but TA as an assemblage has remained. In summary, it has been shown from this review that the role played by TA of public spaces is relevant to the city's urban dynamics with potentially positive benefits contributing to the social landscape. The resilience of an urban environment is strongly linked to the processes of the social landscape of the city (Childers et al., 2014). Moreover, TA is an indicator of urban social sustainability to study and assess the urban landscape at the scale of the public space (2018).

Analysing Assemblages

This exploratory study was carried out adopting a case-study approach. It analyses the legal framwork assemblage through secondary sources, examining laws and

Fig. 7 Moneda St. In Mexico City Centre (Autoridad del Centro Historico, 2011)

regulations governing the use of the street in MCC. The utilised method is called document analysis which is commonly used as a systematic procedure for reviewing and evaluating either printed or digital documents (Bowen, 2009). TA and urban design assemblage analysis were carried out on site, selecting Moneda Street located in MCC. This sample (see Fig. 7) was chosen for the following reasons: (i) it is one of the oldest streets in the area, (ii), it has mixed land-use and building functions (Flores, 2016) and (iii) it has been improved by urban design interventions mentioned earlier. In total, 32 photographs were taken along Moneda St; the starting point was in the corner with Seminario St. next to the Zocalo (main square) and the end point was in EJE 6 av. Visual complexity (VC) analysis in conjunction with a TA observation method is one of the more practical ways to study the urban assemblage (Lara-Hernandez, Melis & Caputo, 2017; Rapoport, 2005; Lara-Hernandez, Melis, & Lehmann, 2018; Gutierrez de Velasco Romo & Padilla Lozano, 2012; Salazar Trujillo, 2010). The diversity of TAs is calculated using the Shannon-Weiner diversity index, which the formula comes as follows: $H' = -\sum \text{piln}$ in which $\text{pi} = 1/\log S$ (S = total number of elements). Thus, the higher the value of diversity index, the higher level of TAs. After calculating the diversity value of each street, the results can be compared and show which of the analysed streets have the higher diversity value of TA.

Analysing the Legal Framework

Table 1 summarises the official public information found in Mexican laws and regulations. The discovery of the absence of clarity surrounding the potential activities on the street, and how citizens appropriate such public spaces, accentuates the need for further clarification of the term and the consensus between different laws and regulations.

To summarise, the three points established by the Mexico City government (2013) are: that public spaces must be accessible and usable by every citizen without no distinction or impediment; that a street may have a function other than simply transportation; and that appropriating public spaces and streets is both a right and a duty for Mexican citizens. The law also provides for streets to be clear of obstacles impeding or hindering pedestrian traffic, unless the obstacle has been authorised. Pedestrian movement thus has legal priority, but uses for the street other than transportation are acknowledged. These laws and regulations do not, however, specify what those other functions may be, and we therefore argue that the laws are at best vague and at worst contradictory and that this could affect the social obligations of citizens and how public spaces are used.

Analysing TA

The TA was analised using a technique known as "activity mapping" that has been developed to analyse street uses occurring at the ground floorlevel and physical elements (Francis, 1984). This technique is used to understand the "temporal city" that takes place in the urban space, stressing the process of interactions between people and the built environment. The observed period was from 7–27 May 2018, during weekdays and at the weekend, as well as at three different times during the day, each for a period of two hours. The activities were mapped as an expression of temporary appropriation, showing the exact location of TA occurring in the public space. Table 2 illustrate the observation analysis in detail.

Analysing Visual Complexity

Adapted from an Australian research-based team (Porta & Renne, 2005), the analysis begins by studying the VC of the streetscape which measures the amount of variety in the streetscape. It attemps to describe the richness of the visual tapestry, allowing to study the streetscape tri-dimmentionally. Figure 8 illustrates a series of photographs taken along the centre of the street, 25 m apart. A total of 32 photos were taken along Moneda St. and a naming code was used to identify them starting with the first letter of the Street name followed by the section number. The pictures were categorised into five main groups according to the VC value. In order to ensure that each photo

Table 1 Laws and regulations about the use of streets in Mexico city

Name	Law for the use of streets and public spaces in Mexico City (Gobierno del Distrito Federal, 2013)	Law of the Civic Culture of Mexico City (Asamblea Legislativa del Distrito Federal III Legislatura, 2004)	Law of Mobility of Mexico City (Gobierno del Distrito Federal, 2016)	Transit Regulations of Federal District (Gobierno del Distrito Federal, 2015)	Citizen Manual for the care of City Centre (Gobierno de la Ciudad de México, 2014)
Generalities and Functions	Stabilised the right to use and enjoy the public space, including streets used in a different way rather than for motor-vehicles	It establishes the minimum rules of citizen behaviour	It dictates the laws for mobility purposes in Mexico City	It regulates the use of streets and roads of Mexico City	It is a manual for citizens that use MCC
Specific articles that mention the use of public space	Art 6. The users of public space and streets have the right to I. Use the property of common use, according to its nature and destiny: II. To access public spaces and streets: III. To stay in public spaces and streets; IV. To transit in public spaces and streets	Second Title of the Civic Culture and neighbour's participation. Chapter I. Art 15. The Civic Culture in Mexico City, which guarantees the harmonious coexistence of its inhabitants, is based on the fulfilment of the following citizen duties. VI. To allow the freedom of action of people in public spaces and streets. X. Make proper use of public goods, spaces and services according to their nature and destination	Art 5. Mobility is the right of every person and the community to carry out the effective movement of individuals and goods to access through the different modes of transportation recognized in the Law, a mobility system that conforms to the hierarchy and principles that are established in this order, to meet their needs and full development. In any case, the object of the mobility will be the person	Art. 1. The purpose of these regulations is to regulate the circulation of pedestrians and vehicles on public roads and road safety in Mexico City. Art. 6.- The Public Administration will provide the necessary means for people to freely choose the way to move in order to access the goods, services and opportunities offered by the City. The level of vulnerability of users, the externalities generated by each mode of transport and their contribution to productivity will be considered for the establishment of public policy in this area. Priority will be given to the use of road space and the distribution of budgetary resources will be assessed according to the following hierarchy of mobility: I. Pedestrians	Chapter 7. Use and conservation of public spaces. Conditions to realise cultural activities. First, permission must be granted by the SSPDF (Secretary of Public Safety of Mexico City), the SGDF (Mexico City Council) and Autoridad del Centro Historico.
	Art 7. The duties of the users are: I. To use the public space and streets according to their nature and destination: II. To exercise their rights without disturbing the order and public peace, or affecting the developing continuity of other inhabitants' activities.		Art 13. For the fulfilment of the present Law and the ordinances that emanate from it, Public Security will have the following attributions: III. Maintain within the scope of its powers, that the road is free of obstacles and elements that impede, hinder or impede vehicle and pedestrian traffic, except in only authorized cases, in which case, as far as possible, they should not obstruct the accesses for persons with disabilities		

Table 2 Observation analysis

Days of observation	Time of observation	Observation technique	Diversity analysis
Monday and Sunday	1. 7:00–9:00 am. The first period of observation is selected according to "rush hour" in MCC. Offices, shops and schools usually start their operation at this time of the day	Three-Round observations of two-hour periods were carried out each day. Each two hour period consisted of 2 rounds of 15 min walking (to and fro, totalling 30 min in each hour); i.e., 4 rounds per 2-hour slot. These rounds of observations were conducted in 6 snapshots in two 15 min walking snap-shots, per hour, per area in a day of observation	Shannon-Weiner diversity index $H' = -\sum piln$
	2. From 2.00–4.00 pm: The second period will be lunch time. Many people use this time to take a break, to go out, and have lunch, therefore the chances of observing a diversity of activities in public spaces are higher		
	3. From 6.00–8.00 pm: The third period is when the majority of people finish their daily working routines		

captured the same amount of visual field, a camera was attached to a tripod (eye level). Once the photos were gathered, each of them was analysed to convene the measurements of the sub-indicators such as colour, façade, urban furniture and pavement. The photos were compiled, and a computer program (Photoshop) was used to analyse them precisely. The main disadvantage of the method proposed by Porta and Renne (2005) is that it could be considered subjective since the measurements are based on personal judgments. In an attempt to make it more objective, the measurements were assessed using a software program for the colour indicator while, for the others, a group of five experts (all architects) appraised and categorised the photos separately. The origin and cultural backgrounds of the experts range from Africa, Asia, Europe, America and Oceania. It follows that in terms of Groat and Wang (2013) this selection (five architects) could be considered as a stratified sampling strategy.

Fig. 8 Moneda St. photos, From M1 top-left to M32 bottom-right

The streetscape was analysed by comparing the VC of each photograph (or street section). The concept of 'visual complexity' is essentially multi-dimensional, thus is best evaluated by reference to four different sub-areas: colour, façade, street furniture and pavement.

1. Colour (the number of different colours, brightness, richness and high contrast) by comparing the standard deviation index of the colour histogram of each photo. Colour histograms are frequently used to compare images (Arman, Hsu, & Chiu, 1993; Hampapur, Jain, & Weymouth, 1995; Ogle & Stonebraker, 1995; Pass, Zabih, & Miller, 1998). The photos were classified into five groups, the highest ranking five (86–80 Std Dev), followed by four (79–73), three (72–66), two (65–59) and the lower one (58–52). Figure 9 illustrates the highest and lowest ranked photos respectively.
2. Façade (attractive doors and cornices, attractive height building articulation and details in roof lines, balconies, verandas, and material's variety). This sub-indicator has been calculated based on the personal judgment of five architects from different countries: Mexico, New Zealand, Algeria, Italy and the UK. Each photograph was evaluated three times with reference to a 1–5 scale shown in Fig. 10.

Fig. 9 Colour sub-indicator from highest (left, rating 5) to lowest (right, rating 1)

Fig. 10 Façade sub-indicator from highest (left, rating 5) to lowest (right, rating 1)

Fig. 11 Street furniture sub-indicator from highest (left, rating 5) to lowest (right, rating 1)

3. Street furniture (benches, street-art, alluring light posts, raised planters, etc.). Similar to the previous sub-indicator, it was also judged based on personal judgement from five different architects with reference to a 1–5 scale illustrated in Fig. 11.
4. Street pavement (variety of texture, colour, different material, patterns, and attractive finishing). Similarly to the previous sub-indicator, it was also judged based on personal judgement from five different architects with reference to a 1–5 scale illustrated in Fig. 12.

The output was then illustrated in a radar graph in order to better understand the relations between the components of the urban design assemblage. The latter is a key difference between the method developed by Porta and Renne (2005) in which the VC is calculated as the average of the sub-indicators.

Fig. 12 Street pavement sub-indicator from highest (left, rating 5) to lowest (right, rating 1)

Findings and Discussion

The first set of questions aimed to describe the cultural and legal landscape, as has been outlined previously. Table 3 summarises the results. It indicates the Shannon-Weiner diversity index (SW Div.) value and the VC value for each of the street sections. M8 and M32 show the highest and lowest values of VC respectively (Fig. 13).

Figure 14 exemplifies street section M8 which obtained the highest VC value and TA value of 1.77 SW div. Figure 15 illustrates street section M32 which obtained the lowest VC value and TA value of 0.74 SW div. Figures 16 and 17 illustrate street sections M17 and M31 with the highest (4.63 SW div) and the lowest (0.25 SW div) TA diversity values respectively.

Table 3 Street sections, TA diversity and VC values

Street section	SW Div	Visual complexity	Street section	SW Div	Visual complexity
M1	1.78	4.25	M17	4.63	3.50
M2	1.02	4.00	M18	3.04	3.75
M3	0.62	4.25	M19	3.25	2.75
M4	0.36	4.25	M20	2.54	3.00
M5	2.41	4.25	M21	2.70	2.50
M6	2.45	4.25	M22	2.74	1.25
M7	2.28	4.25	M23	0.77	2.25
M8	1.77	4.50	M24	0.52	2.50
M9	1.68	3.75	M25	1.31	1.75
M10	3.01	4.25	M26	2.85	2.00
M11	2.25	3.50	M27	0.35	3.00
M12	1.69	3.50	M28	0.42	2.75
M13	2.47	3.00	M29	0.41	2.50
M14	3.97	2.25	M30	0.42	2.75
M15	3.96	3.50	M31	0.22	2.50
M16	2.90	2.50	M32	0.74	1.50

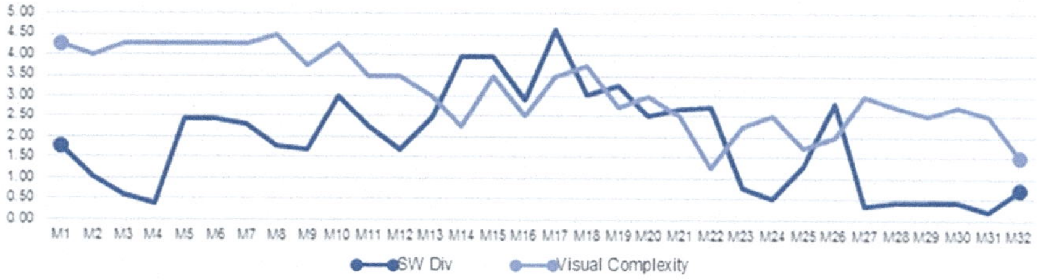

Fig. 13 Linear chart of the street sections and their corresponding TA and VC values

Fig. 14 (right) Street section M8 with the highest VC value, (left) VC chart

Fig. 15 (right) Street section M32 with the lowest VC value, (left) VC chart

Fig. 16 (right) Street section M17 with the highest diversity level of TA of 4.63 SW Div, (left) VC chart

Fig. 17 (right) Street section M31 with the lowest diversity level of TA of 0.22, (left) VC chart

What stands out from the figures is that higher levels of VC as shown in Fig. 14 are not necessarily correlated with higher TA diversity levels values, which are just barely above half of the TA SW div. maximum value observed in Moneda St. The most interesting aspect of these graphs is that they illustrate a correlation between the sub-indicators as an assemblage and TA. For instance, Fig. 17 shows proportionally reduced VC values in comparison with the values shown in Fig. 16 but TA SW div. values are at extremes of the sample (4.63 and 0.22 respectively). This result is somewhat counterintuitive. Taken together, these results suggest that there is a strong relationship between the urban design assemblage (VC) and the diversity of TA in the built-environment. Figure 13 illustrates that the highest TA SW Div values were found in street sections from M13 to M22 ranging from 2.64 to 4.63. The most striking result to emerge from the data is that the VC statistical average of those street Sects. (2.80) is below the statistical average for the whole street Sects. (3.14). If we now turn to the whole TA street performance, these results suggest that higher levels of TA SW Div are found in the central area of the total street sections.

An initial objective of the chapter was to demonstrate how TA in the urban landscape could be analysed using an assemblage thinking approach by illustrating how the interrelations between cultural and regulatory assemblages catalyse TA emergence. The current chapter has illustrated the concept of a highly coded city as a pivotal context in which TA emerges affected to some extent by the relations between the cultural and legal assemblages. Therefore, TA could be explained as an assemblage product of the relations with other assemblages such as the cultural and the legal. These relationships may partly be explained by the fact that even though there is a cultural syncretism (mainly the Aztec and Spanish but not limited to) and the legal and regulatory changes across the centuries which have occurred in MCC, TA is still palpable and evident. The most interesting finding is that the TA of public spaces is mentioned and recognised to some extent by all of the Mexican laws and regulations reviewed so far. Surprisingly, TA is stated as a citizen's right, but also a duty. This finding is contrary to previous results offered by previous authors such as García Espinosa (2005), Hernández Bonilla and Gomez Gomez (2015) and Martinez-Ramirez (2015) who have suggested that regulations have been imposed on public spaces that hinder TA. A possible explanation could be the way in which public space is managed by public or private authorities, which usually segregate or exclude people belonging to minority groups or a lower economic class. Another possible explanation could be attributed to the physical appearance of the built-environment after urban renewals, which could be related to cultural symbols that exclude certain populations or groups within this specific context. A limitation of the chapter is that it only focuses on the laws and regulations for the use of streets (and public spaces) in Mexico City; it does not address how the streets are actually managed as public spaces.

Prior studies have suggested the importance of VC as an urban design indicator and its relation to social sustainability. The findings presented here are contrary to previous studies (Mahdzar, 2008; Francis, 1984; Palaiologou, 2015) which have suggested that higher values of VC indicators are associated with higher levels of social activity (including TA) in the built-environment. These findings raise intriguing questions regarding the nature and extent of the beautification of the urban landscape towards pursuing liveability. The present results are significant in at least two major respects. Firstly, they illustrate that there is a proportional relationship between the values of the VC indicators and TA. Lastly, the higher values of TA SW Div are in the central areas of the street as a whole, meaning that the closer to the square or the avenue the less TA diversity. A possible explanation for this might be because the proximity to a more versatile public space (square) allows more diverse activities to occur, reducing the TA SW Div in the vicinity, while in the case of the avenue it seems that the proximity to heavy car traffic flow discourages TA. A note of caution is due here since there are other urban elements that were not taken into consideration for the study that could affect the TA SW Div such as metro stations. These results provide further support for the hypothesis that first, the higher the level of traffic, the less interaction between people on the street (including TA) (Appleyard, 1980), and secondly, the less availability of open space, the more the street becomes the *space of* contact (Choay, 2001).

These arguments further support the idea of using an assemblage thinking approach to analyse and describe urban phenomena. In addition, the chapter sets out the aim of illustrating the relevance of TA in environmental sustainability and public health. To encourage TA could positively contribute to the urban landscape. Firstly, it promotes the use of outdoor spaces which contribute to the reduction of building energy consumption; secondly, because it is an outdoor activity, it promotes physical and mental health. It can thus be suggested that TA has a positive impact on the urban environment and population. This is an important issue for future research.

Conclusion

The first section of the chapter described the assemblage thinking approach in which assemblages emerge between the interactions of their parts. This chapter has shown that TA is an assemblage that emerges as a product of other assemblages. We agree with DeLanda (2016) by arguing that assemblage theory is extremely resilient because it can be implemented at larger spatial scales and longer temporal scales too (as in the analysis presented here). In conclusion, the present research reinforces the idea that assemblage theory can be used as a theoretical framework for investigating urban-social phenomena. The findings of this research provide insights for the understanding of how the flows and the relationship between components within the urban landscape make emergence assemblages such as TA. This research also contributes to the existing knowledge of TA by offering an assemblage thinking approach as an instrument for studying the urbanity at the scale of the public space. Research in the fields of the built-environment (architecture, landscape architecture, urban planning and urban design), could provide further knowledge on the design settings (design features) of the built-environment (another assemblage) related to TA in any determinate context. In addition, the contribution to other fields of research (sociology, environmental psychology, urban ecology) could help us to shed light and increase certainty in this matter.

Looking at the scope for further interdisciplinary work on TA, one possibly valuable research approach might explore the potentially positive contribution of TA on urban energy consumption. In 2015 the building sector consumed roughly 84.6 EJ globally, accounting for 29% of global final energy consumption (24% for residential and 8% for commercial) (IEA, 2017b) and generated 30% of energy-related CO2 emissions (IEA, 2015). One EJ is fairly close to the electrical power consumption of the world in a given 11-hour period in 2015. The energy-spending on heating and cooling in total building energy consumption ranges between 18 and 73%—the highest being in developing countries located in tropical and subtropical climates by commercial buildings (Ürge-vorsatz et al., 2015). The IEA (2017a) makes a series of recommendations in order to counter this trend such as strengthening and enforcing building energy policies across all countries to prevent the lock-in of long-lived, inefficient building investments, the implementation of educational programs, training and capacity building, and better building of energy data. Another potential extension of our work would be to look at the implications of TA for promoting more activities outdoors rather than indoors which, would have significant benefits for public health. Physical inactivity is a major risk factor associated with non-communicable diseases such as type 2 diabetes (Jeon et al. 2007), breast and colon cancer (Friedenreich, 2010) and coronary heart disease (Sattelmair et al. 2011). Physical inactivity was estimated to be responsible for 5.3 million premature deaths worldwide in 2008 (Lee et al. 2012, 2013). TA of the streetscape by definition occurs outdoors and is opposed to what advocates of health sciences categorise as physical inactivity. Thus, we suggest TA could have a positive impact on public health since it promotes outdoor activities. However, more studies and evidence are required on this matter.

A limitation of this study is that with regard to the urban dimension, it was centred at the streetscape level. A natural progression of this work would be to analyse and study other streets with similar urban conditions, or to focus on TA on different urban scales such as neighbourhoods, districts and cities, and look at the implications of TA for the built environment. Despite its exploratory nature, this study offers some insight into the beneficial aspects of TA for the urban agenda. The findings of this study have a number of practical implications. For instance, it could inform practitioners, professionals and decision makers dealing with urban design interventions aiming to create more inclusive and lively public spaces in central areas of the city and more specifically in UNESCO world heritage city centres. In addition, it could contribute to the development and improvement of guidelines provided by local and international institutions aiming to a better practice. Lastly, it provides a methodology that allows a quick assessment of the street as public space in terms of inclusion.

References

Anderson, B., Kearnes, M., McFarlane, C., & Swanton, D. (2012). On assemblages and geography. *Dialogues in Human Geography*, *2*(2), 171–189. [Online] Available from: https://doi.org/10.1177/2043820612449261.

Appleyard, D. (1980). Livable streets: Protected neighborhoods? *The ANNALS of the American Academy of Political and Social Science*, *451*(1), 106–117. [Online] Available from: https://doi.org/10.1177/000271628045100111.

Araya Diaz, D. F. (2016). Propiedad y apropiación. In: Contested Cities. 2016 Madrid, España. p.

Arman, F., Hsu, A., & Chiu, M.-Y. (1993). Image processing on compressed data for large video databases. In *Proceedings of the First ACM International Conference on Multimedia, ACM* (pp. 267–272).

Asamblea Legislativa del Distrito Federal III Legislatura. (2004). Ley de Cultura Cívica del Distrito Federal. Gobierno Del Distrito Federal, pp.1–29. [Online] Available from: http://www.aldf.gob.mx/archivo-0f05874fac7a0a4b94b9935dd0998eae.pdf.

Autoridad del Centro Historico. (2011). *Plan Integral de Manejo del Centro Historico de la Ciudad de Mexico (2011–2016)*. Mexico, DF.

Bay, J. H. P., & Lehmann, S. (2017). *Growing compact: Urban form, density and sustainability*. New York, USA: Routledge, Taylor & Francis Group. [Online] Available from: https://doi.org/10.4324/9781315563831.

Blanco, J., Bosoer, L., & Apaolaza, R. (2014). Movilidad, apropiación y uso del territorio: Una aproximación a partir del caso de Buenos Aires. Scripta Nova. Revista electrónica de geografia y ciencias sociales. XVIII (493).

Bonfil Batalla, G. (2004). Pensar nuestra cultura. Dialogos de accion, primera etapa. pp. 117–134.

Bowen, G. A. (2009). Document analysis as a qualitative research method. *Qualitative Research Journal*, *9*(2), 27–40. [Online] Available from: https://doi.org/10.3316/QRJ0902027.

Carrera Stampa, M. (2002). Tenochtitlan (Ciudad de México), Evolución (pp. 1325–2012). [Online]. Foro Amistad. Available from: http://www.foroamistad.org/otros-temas/tenochtitlan-(ciudad-de-mexico)-evolucion-1325-2012/. Accessed 29 August 2017.

Childers, D. L., Pickett, S. T. A., Grove, J. M., Ogden, L., et al. (2014). Advancing urban sustainability theory and action: Challenges and opportunities. *Landscape and Urban Planning*. *125*, 320–328. [Online] Available from: https://doi.org/10.1016/j.landurbplan.2014.01.022.

Choay, F. (2001). Historic heritage and the contemporary culture industry. *The invention of the historic monument* (pp. 138–163). Cambridge University Press.

Crossa, V. (2009). Resisting the Entrepreneurial City: Street Vendors' Struggle in Mexico City's Historic Center. *International Journal of Urban and Regional Research, 33*(1), 43–63. [Online] Available from: https://doi.org/10.1111/j.1468-2427.2008.00823.x.

Crossa, V. (2016). Searching for order on the 'dirty' streets of the city. Displacement of street vendors in colonial and postcolonial Mexico City. In: Contested Cities. 2016 Madrid, España. p.

DeLanda, M. (2006). *A new phylosophy of society*. London, UK: Continuum.

DeLanda, M. (2011). *Philosophy and simulation: The emergence of synthetic reason*. London, UK: Continuum International Publishing Group.

DeLanda, M. (2016). *Assemblage Theory*. Speculative Realism. Edinburgh, UK: Edinburgh University Press.
de la Torre, M. I. (2015). Espacio público y colectivo social. *Nova Scientia, 14*(7), 495–510.
Deleuze, G., & Guattari, F. (1989). A thousand plateaus: Capitalism and schizophrenia. In: Brian Massumi (Ed.). Minneapolis, USA: University of Minnesota Press. [Online] Available from: https://doi.org/10.2307/203963.
Deleuze, G., & Parnet, C. (1977). *Deleuze and Claire Parnet. Dialogues*. Paris: Flammarion.
Devlin, R. T. (2017). Asking 'Third World Questions' of First World informality: Using Southern theory to parse needs from desires in an analysis of informal urbanism of the global North. *Planning Theory*, 1–20. [Online] Available from: https://doi.org/10.1177/1473095217737347.
Díaz Larrañaga, N., Grassi, L., & Mainini, C. (2011). Socialidad: Los modos de apropiación del espacio público. *Question 1*(29), 12. [Online] Available from: http://perio.unlp.edu.ar/ojs/index.php/question/article/view/572/849.
Dovey, K. (2012). Informal urbanism and complex adaptive assemblage. *International Development Planning Review, 34*(4), 349–368. [Online] Available from: https://doi.org/10.3828/idpr.2012.23.
Dovey, K., & Pafka, E. (2014). The urban density assemblage: Modelling multiple measures. *Urban Design International, 19*(1), 66–76. [Online] Available from: https://doi.org/10.1057/udi.2013.13.
Dovey, K., Rao, F., & Pafka, E. (2018). Agglomeration and assemblage: Deterritorialising urban theory. *Urban Studies, 55*(2), 263–273. [Online] Available from: https://doi.org/10.1177/0042098017711650.
Drummond, L. B. W. (2000). Street scenes: Practices of public and private space in Urban vietnam. *Urban Studies, 37*(12), 2377–2391. [Online] Available from: https://doi.org/10.1080/00420980020002850.
Ekawati, S. A. (2015). Children—friendly streets as urban playgrounds. *Procedia—Social and Behavioral Sciences, 179*, 94–108. [Online] Available from: https://doi.org/10.1016/j.sbspro.2015.02.413.
Eliot, T. S. (1949). *Notes towards the definition of culture*. [Online] Available from: https://books.google.co.uk/books?id=QYR9FNyKMmwC.
Fleming, R. (1986). Effects of the physical environment on social support and the stress response. Uniformed Services University of Health Sciences. [Online] Available from: http://search.proquest.com/docview/303460175?accountid=14553; http://openurl.library.uiuc.edu/sfxlcl3?url_ver=Z39.88-2004&rft_val_fmt=info:ofi/fmt:kev:mtx:dissertation&genre=dissertations+&+theses&sid=ProQ:ProQuest+Dissertations+&+Theses+Full+Text&atitle=.
Flores, A. (2016). Moneda. Nueva Guia del Centro Historico. [Online] Available from: http://www.guiadelcentrohistorico.mx/kmcero/callejero/moneda. Accessed 2 November 2016.
Flores Arias, V. (2015). El Centro Histórico de la Ciudad de México y sus retos en el siglo XXI. Planur-e. [Online] Available from: http://www.planur-e.es/articulos/ver/el-centro-hist-rico-de-la-ciudad-de-m-xico-y-sus-retos-en-el-siglo-xxi/completo. Accessed 1 May 2016.
Francis, M. (1984). Mapping downtown activity. *Journal of Architectural and Planning Research, 1*(1), 21–35. [Online] Available from: http://www.jstor.org/stable/43028671.
Friedenreich, C. M. (2010). Physical activity and breast cancer: Review of the epidemiologic evidence and biologic mechanisms. In: H. J. Senn & F. Otto (Eds.). *Clinical Cancer Prevention. Recent Results in Cancer Research* [Online]. Berlin, Germany: Springer, Berlin, Heidelberg. p. Available from: https://doi.org/10.1007/978-3-642-10858-7_11.
Fyfe, N. R., & Kenny, J. T. (2005). The urban geography reader. [Online] Available from: http://www.loc.gov/catdir/toc/ecip0421/2004018547.html; http://www.loc.gov/catdir/enhancements/fy0650/2004018547-d.html.
Garcia, E. J. (2013). *The application of eological resilience to urban landscapes*. VIctoria University of Wellington.
García Espinosa, S. (2005). Centros históricos ¿herencia del pasado o construcción del presente? Agentes detonadores de un nuevo esquema de ciudad. Scripta Nova. Revista electrónica de geografia y ciencias sociales. IX (39), 1–11.

Garcia Ramon, M. D., Ortiz, A., & Prats, M. (2004). Urban planning, gender and the use of public space in a peripherial neighbourhood of Barcelona. *Cities*, *21*(3), 215–223. [Online] Available from: https://doi.org/10.1016/j.cities.2004.03.006.

Gobierno del Distrito Federal. (2013). Ley para el uso de las vías y los espacios públicos del Distrito Federal. Diario Oficial de la Federación. [Online] Available from: https://doi.org/10.1017/CBO9781107415324.004.

Gobierno del Distrito Federal. (2015). Reglamento de Transito del Distrito Federal (p. 124). [Online] Available from: http://www2.df.gob.mx/virtual/tlahuac/transparencia/fraccion1/REGLAMENTOS/REGLAMENTO DE TRANSITO DEL DF.pdf.

Gobierno del Distrito Federal. (2016). Ley de Movilidad del Distrito Federal (pp.1–101). [Online]. Available from: http://www.paot.org.mx/centro/leyes/df/pdf/2016/LEY_MOVILIDAD_DISTRITO_FEDERAL_07_10_2016.pdf.

Gobierno de la Ciudad de México. (2014). Manual ciudadano para el cuidado del centro histórico. Fideicomiso Centro Histórico de la Ciudad de México (p. 95.). [Online] Available from: http://www.autoridadcentrohistorico.cdmx.gob.mx/storage/app/uploads/public/58a/490/da3/58a490da3b8ca502567977.pdf.

Gobierno de la Ciudad de Mexico. (2016). Autoridad del Centro Historico. [Online] Available from: http://www.autoridadcentrohistorico.df.gob.mx. Accessed 1 April 2016.

Gomez Flores, L. (2013). Comercio informal en el DF salvavidas y conflicto. La Jornada. [Online] 21 September, 1–4. Available from: https://jornada.unam.mx.

Graumann, C. F. (1976). The concept of appropriation (aneignung) and modes of appropriation of space, *6*(2), 301–313. [Online] Available from: http://iaps.scix.net/cgi-bin/works/Show?iaps_00_1976_009.

Graumann, C. (1983). On multiple identities. *International Social Science Journal, 35*(96), 309–321. [Online] Available from: http://eds.b.ebscohost.com/eds/pdfviewer/pdfviewer?sid=4e7bdb42-733e-4fbb-98b0-3082b60477c5%40sessionmgr105&vid=7&hid=112.

Groat, L., & Wang, D. (2013). *Architectural research methods*. Second. New York, US: Wiley.

Groth, J., & Corijn, E. (2005). Reclaiming urbanity: Indeterminate spaces, informal actors and urban agenda setting. *Urban Studies*, *42*(3), 503–526. [Online] Available from: https://doi.org/10.1080/00420980500035436.

Guattari, F., & Rolnik, S. (2005). *Micropolitica Cartografias del deseo*. Madrid: España, Traficantes de sueños.

Gutierrez de Velasco Romo, E. G., & Padilla Lozano, F. (2012). Apropiación y uso del espacio público en el centro de la ciudad de Aguascalientes. *Investigación y Ciencia, 20*(56), 33–43.

Haan, H. De. (2005). Social and material appropriation of neighborhood space: Collective space and resistance in a dutch Urban Community. October, pp. 1–122.

Hampapur, A., Jain, R., & Weymouth, T. E. (1995). Production model based digital video segmentation. *Multimedia Tools and Applications*, *1*(1), 9–46.

Hawkes, T. (1977). *Structuralism and Semiotics*. New York, USA: Routledge, Taylor & Francis Group.

Hernández Bonilla, M., & Gómez Gómez, S. (2015). Conflicto espacial, exclusión y espacio público en la centralidad urbana de Xalapa, Veracruz. *Revista Brasileira de Gestão Urbana*, *7*(3), 281–294. [Online] Available from: https://doi.org/10.1590/2175-3369.007.003.AO04.

Hernandez Mendo, A., & Morales Sanchez, V. (2008). Una revisión teórica: ocio, tiempo libre y animación sociocultural. Efdeportes.com, (27), 9–12. [Online] Available from: http://www.efdeportes.com/efd127/una-revision-teorica-ocio-tiempo-libre-y-animacion-sociocultural.htm.

Hubbard, P., Kitchin, R., Bartley, B., & Fuller, D. (2002). *Thinking geographically*. London: UK, Continuum.

IEA. (2015). Key World Energy 2015. International Energy Agency.

IEA. (2017a). Energy Technology Perspectives 2017. IEA. [Online] Available from: https://doi.org/10.1787/energy_tech-2014-en.

IEA. (2017b). Key World Energy Statistics 2017. International Energy Agency. [Online] Available from: https://doi.org/10.1017/CBO9781107415324.004.

Jeon, C. Y., Lokken, R. P., Hu, F. B., & Van Dam, R. M. (2007). Physical activity of moderate intensity and risk of type 2 diabetes: A systematic review. *Diabetes Care*, *30*(3), 744–752. [Online] Available from: https://doi.org/10.2337/dc06-1842.

Jessop, B., Brenner, N., & Jones, M. S. (2008). Theorizing sociospatial relations. Environment and planning D. *Society and Space*, *26*(3), 389–401. [Online] Available from: https://doi.org/10.1068/d9107.

Keller, A. H. (2006). A road by any other name: Trails, paths, and roads in Maya language and thought. In: Landscapes of Movement Symposium (p. 29).

Kent, S. (1990). *Domestic architecture and the use of space*. Cambridge: UK, Cambridge University Press.

Kim, A. M. (2013). *Sidewalk city. Chicago*. US, The University of Chicago Press. [Online] Available from: https://doi.org/10.7208/chicago/9780226119366.001.0001.

Kramer, A. (2019). Philosophical materialism: A study of early modern literature and contemporary theory. The alliance for networking visual culture. [Online] Available from: http://scalar.usc.edu/works/material_philosophy/flat-ontology. Accessed 28 April 2019.

Kyle, G. T., Jun, J., & Absher, J. D. (2014). Repositioning identity in conceptualizations of human–place bonding. *Environment and Behavior, 46(8)*, 1018–1043. [Online] Available from: https://doi.org/10.1177/0013916513488783.

Lara-Hernandez, J. A., & Melis, A. (2018). Understanding the temporary appropriation in relationship to social sustainability. *Sustainable Cities and Society. 39C*, 1–14. [Online] Available from: https://doi.org/10.1016/j.scs.2018.03.004.

Lara-Hernandez, J. A., Melis, A., & Caputo, S. (2017). Understanding spatial configuration and temporary appropriation of the street in latin American cities: The case of Mexico city centre. In: H. Boudagh, A. Versaci, F. Trapani, M. Migliore, et al. (Eds.). *Urban Planning and Architectural Design for Sustainable Development* (pp. 153–173). Palermo, Italy: IEREK Springer.

Lara-Hernandez, J. A., Melis, A., & Lehmann, S. (2018). Temporary appropriation and social sustainability: The case of the historic centre of Mexico City. In: *APRU Sustainable Cities and Landscapes*. Hong Kong. [Online] Available from: https://apru.org/.

Lee, I. M. M., Bauman, A. E., Blair, S. N., Heath, G. W., et al. (2013). Annual deaths attributable to physical inactivity: Whither the missing 2 million? *The Lancet*, *381*(9871), 992–993. [Online] Available from: https://doi.org/10.1016/S0140-6736(13)60705-9.

Lee, I. M., Shiroma, E. J., Lobelo, F., Puska, P., et al. (2012). Impact of physical inactivity on the world's major non-communicable diseases. *The Lancet*, *380*(9838), 219–229. [Online] Available from: https://doi.org/10.1016/S0140-6736(12)61031-9.

Lefebvre, H. (1971). De lo rural a lo urbano. J. Gonzalez-Pueyo (Ed.). Barcelona, España: Ediciones Peninsula.

Lehmann, S. (2009). Back to the city: Strategies for informal urban interventions: Collaboration between artists and architects. Hatje Cantz.

Leon Portilla, M. (1995). *Antologia de Tehotihuacan a los Aztecas*. Fuentes e interpretaciones históricas. First re-i. Mexico, DF: Universidad Nacional Autonoma de Mexico.

Lima, J. J. (2001). Socio-spatial segregation and urban form: Belem at the end of the 1990's. *Geoforum*, *32*(4), 493–507. [Online] Available from: https://doi.org/10.1016/S0016-7185(01)00019-7.

Loukaitou-Sideris, A., & Ehrenfeucht, R. R. (2009). *Sidewalks conflict and negotiation over public space*. USA: Massachusetts Institute of Technology.

MacDonald, R., & Shildrick, T. (2007). Street corner society: Leisure careers, youth (sub)culture and social exclusion. *Leisure Studies*, *26*(3), 339–355. [Online] Available from: https://doi.org/10.1080/02614360600834826.

Madanipour, A. (1996). *Design of urban space. An inquiry into socio-spatial process*. New York, US: Wiley.

Mahdzar, S. S. B. S. (2008). *Sociability vs accessibility urban street life*. United Kingdom: University College London.

Marston, S. A., Jones, J. P., Woodward, K., Paul, J., et al. (2005). Human geography without scale. *The Royal Geographical Society, 30* (4), 416–432. [Online] Available from: https://www.jstor.org/stable/3804505.

Martínez, E. (2014). Configuración urbana, hábitat y apropiación del espacio. Scripta Nova. Revista electrónica de geografia y ciencias sociales. XVIII (493), p. 20.

Martinez-Ramirez, U. (2015). Proceso de Gentrificación y Dezplazamiento en el espacio público del Centro Histórico de la Ciudad de México. [Online] Available from: https://doi.org/WPCC-15006.

Marx, C., & Kelling, E. (2018). Knowing urban informalities. *Urban Studies, 1*(16). [Online] Available from: https://doi.org/10.1177/0042098018770848.

McFarlane, C. (2011). The city as assemblage: Dwelling and urban space. *Environment and Planning D. Society and Space, 29*(4), 649–671. [Online] Available from: https://doi.org/10.1068/d4710.

Monnet, J. (1995). Usos e imágenes del centro histórico de la Ciudad de México. Centro de estudios mexicanos y centroamericanos (Ed.).

Monnet, J. (1996). Espacio público, comercio y urbanidad en Francia. *México y Estados Unidos. ALTERIDADES., 6*(11), 11–25.

Moudon, A. V. (1991). *Public streets for public use*. New York, US: Columbia University Press.

Mouffe, C. (2012). Space, hegemony and radical critique. *Spatial Politics: Essays for Doreen Massey*, 19–31. [Online] Available from: https://doi.org/10.1002/9781118278857.ch1.

Mouffe, C. (2014). Agonística. Pensar el mundo políticamente. In: Laclau S. (Ed.). Mexico, DF, Fondo de Cultura Económica de Argentina.

Müller, M., & Schurr, C. (2016). Assemblage thinking and actor-network theory: conjunctions, disjunctions, cross-fertilisations. *Transactions of the Institute of British Geographers, 41*(3), 217–229. [Online] Available from: https://doi.org/10.1111/tran.12117.

Nelson, H. J. (1963). Townscapes of Mexico: An example of the regional variation of townscapes. *Economic Geography, 39*(1), 74–83. [Online] Available from: http://www.jstor.org/stable/142496.

Nivon Bolan, E., & Sanchez Bonilla, D. (2014). La gestión del Centro Histórico de la Ciudad de Mexico: 1980-2012. In: L. Durán, E. K. Garcés, & M. Lacarrieu (Eds.). *Habitar el Patrimonio Nuevos aportes al debate desde America Latina* (pp. 48–65). Quito, Ecuador.

Ogle, V. E., & Stonebraker, M. (1995). Chabot: Retrieval from a relational database of images. *Computer, 28*(9), 40–48.

Palaiologou, G. (2015). *Between buildings and streets*. University College London.

Pass, G., Zabih, R., & Miller, J. (1998). Comparing images using color coherence vectors. In *Proceedings of the Fourth ACM International Conference on Multimedia (MULTIMEDIA '96)* (pp. 1–14). [Online] Available from: https://doi.org/10.1145/244130.244148.

Patiño, D. (2016). Se acabó el Distrito Federal: ¡Bienvenida, Ciudad de México! Expansion. [Online] 29 January, 2016–2017. Available from: http://expansion.mx/economia/2016/01/29/pena-nieto-promulga-la-reforma-politica-de-la-cdmx.

Pol Urrútia, E. (2002). El modelo dual de la apropiación del espacio en Mira, R., Sabucedo, JM y Romay, J. Psicología y Medio Ambiente. Aspectos psicosociales, educativos y metodológicos, pp. 123–132.

Porta, S., & Renne, J. L. (2005). Linking urban design to sustainability: Formal indicators of social urban sustainability field research in Perth, Western Australia. *Urban Design International, 10*(1), 51–64. [Online] Available from: https://doi.org/10.1057/palgrave.udi.9000136.

Portal, M. A. (2009). Memoria y territorio: La sacralización simbólica del espacio urbano. In: La Historia Oral y la multidisciplinariedad: Retos y perspectivas (pp. 1–9). Colima, Mexico: Departamento de Antropología UAM-1.

Purcell, M. (2002). Excavating lefebvre: The right to the city and its urban politics of the inhabitant. *GeoJournal, 58*, 99–108. [Online] Available from: https://doi.org/10.1023/B:GEJO.0000010829.62237.8f.

Raman, S. (2010). Designing a liveable compact city physical forms of city and social life in urban neighbourhoods. *Built Environment, 36*(1), 63–80. [Online] Available from: https://doi.org/10.2148/benv.36.1.63.

Ramirez Kuri, P. (2010). La ciudad y los nuevos procesos urbanos. *Sociología Urbana*, 163–187.
Ramirez Kuri, P. (2015). Espacio público, ¿espacio de todos? Reflexiones desde la ciudad de México. *Revista mexicana de sociologia*, *1*(77), 7–36.
Ramirez-Lovering, D. (2008). *Opportunistic urbanism*. Melbourne, Australia: RMIT Press.
Rapoport, A. (1998). Using 'Culture' in housing design. In: Housing and society. 1998 Seul, Korea. pp. 1–20.
Rapoport, A. (2005). *Culture, architecture, and design*. Chicago: Locke Science Publisher.
Relph, E. (1976). *Place and placelessness*, 2nd edn. Pion Limited.
Salazar Trujillo, J. H. (2010). *Uso y apropriación de los espacios públicos*. Medellin, Colombia, Impregón S: A. Medellín.
Sattelmair, J., Pertman, J., Ding, E. L., Kohl, H. W., et al. (2011). Dose response between physical activity and risk of coronary heart disease: A meta-analysis. *Circulation*, *124*(7), 789–795. [Online] Available from: https://doi.org/10.1161/CIRCULATIONAHA.110.010710.
Seaman, J., & McLaughlin, S. (2014). The importance of outdoor activity and place attachment to adolescent development in Coös County, New Hampshire. Building knowledge for families and communities, 208(37). [Online] Available from: http://scholars.unh.edu/carsey/208/.
Seghezzo, L. (2009). The five dimensions of sustainability. *Environmental Politics*, *18*(4), 539–556. [Online] Available from: https://doi.org/10.1080/09644010903063669.
Simpson, P. (2011). Street performance and the city: Public space, sociality, and intervening in the everyday. *Space and Culture*, *14*(4), 415–430. [Online] Available from: https://doi.org/10.1177/1206331211412270.
Stanislawski, D. (1947). Early Spanish town planning in the new world. *American Geographical Society*, *37*(1), 94–105.
Suárez Pareyón, A. (2004). El centro histórico de la Ciudad de México al inicio del siglo XXI. *Revista invi*, *19*(51), 75–95.
Tylor, E. B. (1871). *Primitive Culture: researches into the development of mythology, philosophy, religion, art, and custom*. London, UK: John Murray.
UNESCO. (2013). World Heritage List. [Online] Available from: http://whc.unesco.org/en/list/. Accessed 29 September 2015.
UNESCO. (2017). UNESCO World Heritage Centre—World Heritage List. [Online] Available from: http://whc.unesco.org/en/list/&type=cultural&order=region. Accessed 16 October 2017.
United Nations. (2014). World's population increasingly urban with more than half living in urban areas. Department of Economic and Social Affairs. [Online] Available from: https://www.un.org/en/development/desa/news/population/world-urbanizationprospects- 2014.html. Accessed 29 September 2015.
Ürge-vorsatz, D., Cabeza, L. F., Serrano, S., & Barreneche, C. (2015). Heating and cooling energy trends and drivers in buildings. *Renewable and Sustainable Energy Reviews*, *41*, 85–98. [Online] Available from: http://dx.doi.org/10.1016/j.rser.2014.08.039.
Vázquez, C. A., & Tapia Quevedo, J. (2011). Espacio público e informalidad: El caso del programa de apoyo para la reubicación del comercio popular en la ciudad de México. *Revista Ujaen*, 139–158.
Vidal Moranta, T., & Pol Urrútia, E. (2005). La apropiación del espacio: Una propuesta teórica para comprender la vinculación entre las personas y los lugares. *Anuario de Psicologia*, *36*(3), 281–297. [Online] Available from: https://doi.org/10.1006/jevp.2000.0185.
Webster, D., & Sanders, W. (2001). La antigua ciudad mesoamericana: Teoría y concepto. *Urbanismo En Las Sociedades Antiguas*, 43–64. [Online] Available from: http://dialnet.unirioja.es/servlet/articulo?codigo=2776058.
Whyte, W. H. (1980). *The social life of small urban spaces*. [Online] Available from: https://doi.org/10.1177/089124168201000411.

Yory García, C. M. (2003). Topofilia, ciudad, territorio: Una estrategia pedagógica de desarrollo urbano participtivo con dimensión sustentable para las grandes metrópolis de América Latina en el contexto de la globalización: 'El caso de la ciudad de Bogotá'. Universidad Complutense de Madrid. [Online] Available from: http://biblioteca.ucm.es/tesis/ghi/ucm-t26725.pdf.

Yory, M. C. (2011). El concepto de topofilia entendido como Teoría del Lugar. *Revista Barrio Taller*, 1–17.

Temporary Appropriation and Informality

Jose Antonio Lara-Hernandez, Alessandro Melis and Claire M. Coulter

Abstract This paper seeks to explore the definitions and uses of both informality and temporary appropriation in an urban context. Temporary appropriation is usually overlooked and misinterpreted as a form of informal practice. By separating temporary appropriation from informality, we have been able to create a simpler model for understanding urban informality, whilst at the same time crystallising temporary appropriation as a concept in its own right. We challenge the existing definitions of informality as applied in studies concerning informal practices in the Global North and argue that some understandings of informal behaviour in urban settings could be better understood when viewed as forms of temporary appropriation. We provide a categorisation of temporary appropriation by providing examples from different cities and cultures and show that there is a definite need for a deeper understanding of temporary appropriation and its relationship with the urban informality of any socio-cultural landscape.

Keywords Informality · Temporary appropriation · Urban informality

Introduction

The urban sciences are increasingly concerned with informality and what it can contribute to the global urban agenda (Georgieva, 2018; Kuppinger, 2019; Moatasim, 2018; Pasquetti & Picker, 2017; Sarmiento & Tilly, 2018). Yet informality is difficult to define, and temporary appropriation (TA) is often seen as merely another form of informal activity. Among the challenges faced by researchers is first to define informality within the urban agenda, and second to identify and categorise the different forms of informality in cities (Devlin, 2017; Marx & Kelling, 2018). Following on from this, finding out where informality and TA intersect and overlap is a challenge in itself.

Looking at informality, Holston (2008) states, "…in the process of building their residential spaces, [informal actors] not only construct a vast new city, but on that

J. A. Lara-Hernandez (✉) · A. Melis · C. M. Coulter
University of Portsmouth, Portsmouth, UK
e-mail: Jose.LaraHernandez@myport.ac.uk

A. Melis et al. (eds.), *Temporary Appropriation in Cities*,
https://doi.org/10.1007/978-3-030-32120-8_4

basis, also constitute it as a polis with a different order of citizen- ship" (2008: p. 311). Some scholars (Caldeira, 1996; Lucia Bayer, 2016; Roy, 2012; Roy & Alsayyad, 2004a, b) have described informality as urban practices outside of the legal and economic framework, others (Kinder, 2014; Mayer, 2007) have associated the concept as being part of bottom-up counter strategies against liberal policies, and others as a tool through which urban residents produce the city (Lydon et al., 2012, 2014). Different though the approaches may be, they all refer indirectly to unplanned, spontaneous activities in public spaces as informal practices and it is this naming of a phenomenon that allows it to be explored to find out—and possibly empower—what meaning it has for those who live "in it" (Peattie, 1987). Be that as it may, the above definitions of informality give misleading implications of a binary condition.

There is no requirement that TA be informal. It refers essentially to the "borrowing" of public space for a limited time and, while it will be informal more often than not, it is possible to find formalised TA including the licenced busking on the London Underground, the licensing of sidewalk vendors in California and all the pop-up, Christmas, farmers' and other markets. The significance is not in the formality or informality but in the temporary nature of the proceedings.

In this chapter we seek to explore the definitions and uses of both informality and TA in an urban context. It is not the aim of this chapter to discuss the concept of informality within the urban realm, which has been widely discussed elsewhere (Auerbach et al., 2018; Boer et al., 2017; Holston, 2008; Roy & Alsayyad, 2004a, b; Roy, 2009, 2012; Lutzoni, 2016; Sarmiento & Tilly, 2018). This chapter is not looking at informality as a template for expanding informal cities. Our focus is on TA in the shape of informal behaviours on the part of people occupying a public space in a formal city. We will examine the conflation of TA and informality and seek to understand the differences clearly. We contend that TA is more than merely an informal activity subset, and that it is necessary to understand it if we are fully to understand the day-to-day use of public spaces in a variety of urban contexts.

Defining Informality

It is difficult to find a common definition of informality; a definition common to different major languages and cultures can be constructed, but loses some of the nuances present in the different original definitions. The Oxford English Dictionary (2015) defines the informal as (1) not done or made according to a recognized or prescribed form; not observing established procedures or rules; unofficial; irregular and (2) characterized by absence of formality or ceremony; casual, relaxed. Merriam-Webster (2018) similarly refers to the informal as being (1) marked by the absence of formality or ceremony, or (2) characteristic of or appropriate to ordinary, casual, or familiar use. Looking beyond the Anglophone world, it is interesting to compare how informality is defined in other major European languages. The Real Academia Española (2017) defines *informal* in Spanish as (1) the quality of being non-formal and (2) banned action because of its informal condition. In German, Duden (2018) defines *informell* as (1) something without a formal assignment and (2) something lacking formality, or which is non official. Treccani (2018) defines the term *informale* in Italian as something which is non-official, whereas in French, *informel* will be a new addition to the 9th edition of the *Dictionnaire de l'*Académie française (2018), showing how this concept has come to prominence since the previous version of the dictionary was published in 1935. In Polish *nieformalny* is defined as something (1) inconsistent with applicable regulations, and (2) not having an official character (PWN, 2018). Finally in Dutch *informeel* is defined as (1) non-formal, unofficial and (2) provisional, non-binding (Van Dale, 2018). Despite the similarities between the definitions across the different languages, there is no definitive definition. Every definition refers to informality, mainly to unofficial and/or irregular behaviour, a few to actions that are banned and only one to something temporary. While all the languages mentioned here have a European origin, there is no fixed, consistent definition of informality in even a limited geographical area (Devlin, 2017; Holston, 2008; Marx & Kelling, 2018). Informality is generally seen as adaptable, resilient and dynamic, capable of filling voids left by governmental indifference (Cordera, Ramirez Kuri & Zlccardl, 2008; Devlin, 2017). Nevertheless, types of informality are distinguished according to the context in which informality occurs. Devlin (2017), for example, sees less richness in the "Global North's" literature on the subject than in that from the "Global South" and points to a lack of empirically grounded work on informal use of space, an argument supported by Marx and Kelling (2018) who contend that urban theory sees informality as mainly something that helps understand non-western cities. In our view, these opinons reveal a narrow understanding that sees informality as only possible in informal settlements and don't recognise that TA can be an informal activity in a formal city whatever the socio-economic setting.

Urban Informality

Analysis of informality in urban studies has been directed largely at challenges to formal systems. "Urban informality" is usually used when referring to illegal occupation of land occupation by poor and marginalised people (Lutzoni, 2016; Rios, 2014; Roy & Alsayyad, 2004a, b; Roy, 2005), or to people and practices defined by the economic characteristics of their type of employment and main income sources and described as the "informal sector" (Binelli, 2016; Cordera, Ramirez Kuri & Zlccardl, 2008; Crossa, 2016; Duhau & Giglia, 2007). For Bayat (1997), poor people aspire to belong and to lead integrated lives but face social, cultural and economic costs. What the disenfranchised want is redistribution ensuring survival, improved living conditions, and access to public goods, for example by the illegal occupation of land and shelter construction. In the Middle East, the poor use the language of morality or urban politics to justify illegal activity, as do those in Latin America (Sen, 1999).

Looking beyond the Global South, Devlin (2017) claims the existence of two different types of drivers for informal practices, (i) the acts undertaken by the urban poor in order meet basic needs and to survive, and (ii) those engaged in by more well-off residents for convenience, efficiency, or creative expression. The former could be defined as the "informality of need" while the latter as the "informality of desire". The risk inherent here is that Devlin views both forms of informal practice under the same terms, and makes the tacit assumption that all expressions of informal acts are politically neutral, organic or bottom-up, which in turn has the potential to lend endorsement to neoliberal projects of de-regularisation and state retrenchment (Iveson, 2013; Kinder, 2014).

Furthermore, Devlin (2017) categorises studies of informality in three different groups: informality as everyday authenticity, as best practice, and as (Lefebvrian) cry and demand. The first group represents advocates such as De Certeau (1984) and Jacobs (1961) who drew attention to dimension unseen by the planners and designers who were missing the social landscape of the streets. The second group corresponds to advocates such as Deleuze and Guattari (1989). Although their theoretical discussions never refer to informality per se, they criticise formal structures of planning and administration (Bogue, 1991). It is characterised by practitioners encouraging citizens to appropriate the urban realm through creative and informal practices. Lydon et al. (2014) and their Tactical Urbanism strategies are a good example of this group. Lastly, the third group is divided in two sub-groups: DIY (Do-It-Yourself) Urbanism, and the group of studies referring to community gardens in the USA, Canada and European cities. Both sub-groups are devoted to the notion of the commons and the conflict between use-value and exchange-value of urban space described by Lefebvre (Devlin, 2017). These categories of informal use range from housing to leisure activities, sharing a Lefebvrian conceptualisation of the politics of spatial production. Devlin argues that direct actions from informal actors aim to fulfil a collective need expressed through social disobedience against property rights. There is a "cry and demand" from urban residents to undertake planning without the state, which is assumed to be a regressive force captured by and serving capitalist interests (Purcell, 2016).

Marx and Kelling (2018) categorise informality in the field of urban studies differently. They identified three main approaches in how Anglophone researchers have studied the phenomena in the "western world": informality as a condition, as a law and as a currency. Informality as a condition is the most common way of thinking about informality. For instance, UN-Habitat (2009, 2013) defines informality on the basis of the nature of property rights, planning, infrastructure and level of services. Marx and Kelling (2018) criticise this view by arguing that it leads to debates about precision and reliability, rather than exploring the reasons about why and how they are applied. Informality as a law is associated with legal pluralism, which holds that plurality of legal systems co-exist, such as state law, religious law, indigenous law, customary laws and local conventions. In such a plural realm the law of the state is not necessarily the dominant one. Furthermore, the state might not be able to have the capacity to enforce the law (McAuslan, 2006). And so what is taken as formal could be informal and vice versa. A more detailed study is needed to understand the context and the part TA may play. Lara-Hernandez and Melis (2018) support this argument by arguing that the cultural dimension plays a decisive role in how people make use of spaces. Be that as it may, the fact remains that a legal apparatus of some sort continues to operate. Informality in any situation is not necessarily the most important aspect of what is happening. Whether informally or formally, human practices are social constructs and are socio-politically significant. Informality emphasises change, flux, changes and the dynamic while not losing sight of the ambiguous and even contradictory practice of regulating the use of public spaces in cities (Lara-Hernandez, Melis & Coulter, 2018). Informality, then, is the reverse of formality but form and organisation are not necessarily lacking. To take this approach is to acknowledge the significance of place, context and local dynamics in informality rather over neoliberal urges. Rapoport (2005) states that such studies should take into consideration the interplay between people's activities in their built environment, which is strongly linked to culture. For instance, an easy and extreme example would be a house. He provides an example based on housing by comparing the traditional western setting with a tribal one. Figure 1 illustrates how different activities that take place inside house "A" may occur in a widely dispersed system of settings in the case of house "B". Thus, the units to be studied are not the houses themselves but the system of settings within which particular systems of activities take place. This type of conceptualisation makes a huge difference because it enables us to see the striking disparity between the uses of the two different houses and by extension the use of the street, the neighbourhood and so on.

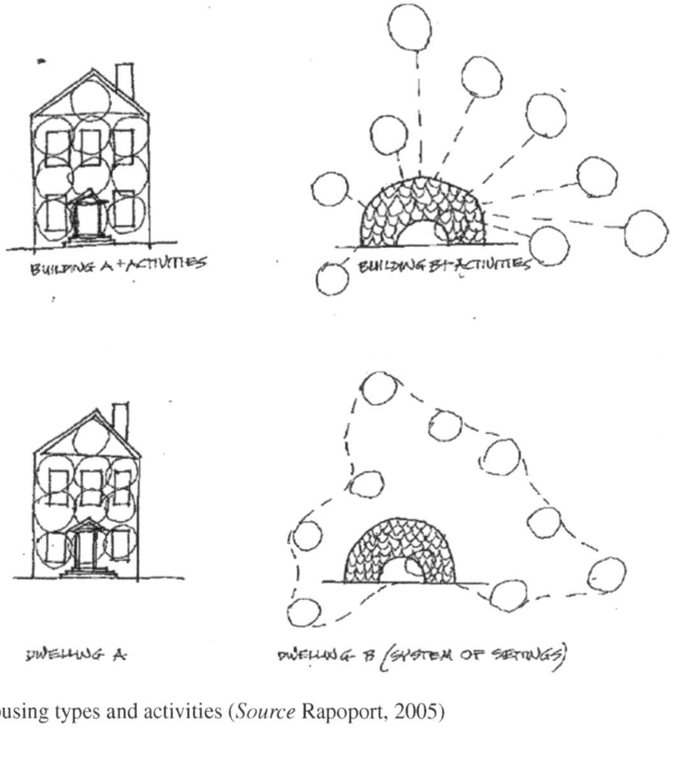

Fig. 1 Housing types and activities (*Source* Rapoport, 2005)

In view of all that has been mentioned so far, one may suppose that what could be described as informality in a given context could really be just a different reflection on local dynamics, and relations between people and their built environment. We would like now to explain how TA can be considered as an expression of these local dynamics occurring in the urban environment, explaining in more depth the nature of TA as an urban phenomenon.

Temporary Appropriation

People have an inborn socio-spatial special need to appropriate the built environment (Graumann, 1976; Yory, 2011). According to Lefebvre (1971), the concept of appropriation is one of the most remarkable results that centuries of philosophical reflections have bequeathed. Appropriation is the pinnacle of a social life; without appropriation our domination of the natural environment would be absurd. Yory (2011) supports this by arguing that for humanity the construction-appropriation of our environment is something inherent and connatural. The concept of appropriation is implicit in his theoretical definition of topophilia as "the act of co-appropriation generated between the man and the world; through which the world becomes the world, at the opening realised by the man within its historic-spatial nature and human becomes human through its spatialisation" (Yory, 2011: p. 45). Whereas humanisation is where the individual internalises a set of fundamental societal defined meanings (Graumann, 1976), appropriation, as an inborn necessity of humans, can be expressed through activities that occur in the urban landscape. Authors such as Pol Urrútia (2002), Purcell (2002) and Yory García (2003) have drawn respectively upon Lefebvre, the dual model and topophilia to use appropriation as a term to explain the theoretical link between people and places.

Public spaces are an essential part of the urban landscape and play host to a range of publicly occurring activities which are of a temporary nature (Madanipour, 1996; Moudon, 1991; Noschis et al., 1978). Appropriation is a broad term, and the majority of the literature on the general theme of appropriation refers to it as a lasting and consistent condition (Ardura Urquiaga, 2014; Díaz & Ortiz, 2003; Ramirez-Lovering, 2008). Less attention has been paid to appropriation that is temporary and includes events and actions in public spaces, is not permanent, or is impromptu (Fonseca Rodriguez, 2015; Lara-Hernandez, Melis & Caputo, 2017).

Cities' worldwide unrelenting growth arouses reflection on the sustainability of development (Lehmann, 2015), and the use of urban spaces. TA is a result of the need to adapt to a cityscape that changes in a way that removes reference points familiar to its populace. Appropriation's temporary nature becomes essential to urban flexibility and may be seen as the right of its citizens as social beings living in urban settings.

Regarding the urban landscape, some authors (Roy, 2009; Roy & Alsayyad, 2004a, b; Ramirez-Lovering, 2008; Lucia Bayer, 2016) have used TA to refer to the informal or illegal use of urban space. While this may be an appropriate response to private land, it raises problems when referring to the use of public space as the owners of public space are the citizens. That this is so may be felt more strongly in Latin America, the Mediterranean and Southeast Asia but TA—though more visible there—is not purely regional. In fact, Fonseca Rodriguez (2015) provides a definition of TA as “the temporary act in which people use public spaces to carry out individual or collective activities other than the purpose that the space was originally designed for,” a definition which is incredibly open to a wide range of activities and settings.

TA therefore lets citizens redefine and reshape urban spaces to meet their own needs instead of simply accepting what is there. Citizens live out their right to the full management and use of the urban environment (Lefebvre, 1992; Purcell, 2002). Graumann (1976) notes that the appropriation of public space is a medium and a goal in order to overcome human alienation, whilst Lara-Hernandez and Melis (2018) argue that TA is the ephemeral and time framed appropriation of the urban environment reflecting an individual, social, and spatial need.

Lara-Hernandez, Melis and Caputo (2017) argue that TA can also be defined as the interaction between citizens and their city expressed through certain kinds of activities occurring in public spaces. Drawing upon the work of Crossa (2009), Drummond (2000) and Monnet (1995), it is possible to identify three groups of activities that can be considered as expressions of TA: activities related to the economy such as work/trade (Crossa, 2016, 2009; de la Torre, 2015; Kim, 2013; Loukaitou-Sideris & Ehrenfeucht, 2009) illustrated in Fig. 2, activities related to leisure (Díaz Larrañaga, Grassi & Mainini, 2011; Drummond, 2000; Hernandez Mendo & Morales Sanchez, 2008; MacDonald & Shildrick, 2007; Mouffe, 2014; Seaman & McLaughlin, 2014; Simpson, 2011) shown in Fig. 3 and activities related to religion, culture and heritage (Portal, 2009; Monnet, 1995) presented in Fig. 4. All three contribute towards maintaining the social urban landscape.

Fig. 2 (left) Woman selling clothes in the street in Hong Kong; (right) Man selling handcraft souvenirs in Auckland’s CBD (*Source* Authors)

Fig. 3 (left) Man exercising in Xian, China; (right) Band playing music in Newcastle, UK (*Source* Authors)

Fig. 4 (left) Altar in the street in Mexico City Centre; (right) Remembrance objects on a bench in Portsmouth, UK (*Source* Authors)

This way of looking at TA makes it easier to distinguish between simply using the space and appropriating it. Seghezzo (2009) calls one of his five dimensions of sustainability "persons" which is not the same as Leary-Owhin (2015)'s definition, after Lefebvre, of TA as temporally bounded and imposed on urban spaces. Not every activity that can take place in a public space is TA. Lara-Hernandez, Melis and Caputo (2017) previously categorised them into three main groups which are economy, leisure and sacralisation (activities related to religious or cultural practices). Table 1 illustrates the three main groups.

Thus far we have explained TA as an urban phenomenon that emerges as an expression of local dynamics and relations within a local context, and as something deeply rooted in our human nature. The section that follows moves on to consider TA in relationship to informality in the urban arena.

Temporary Appropriation and Informality

Previously was argued that TA is activated by a sense of identity and belonging whose nature is predominantly informal. Likewise informality can be seen as a temporary condition which places constraints and restrictions on existing public spaces, structures or norms. As argued at the start of this chapter, the use of the term 'informality' by definition implies the assumption of a binary condition, where activities are assigned to either the informal or the formal, and as we have demonstrated in the section above, this is not always appropriate for activities which as part of TA occupy the grey areas in between.

Bayat (1997) examined the overlap between TA and informality and concluded that informality is not the urban poor's essential preference but an alternative to restraints imposed by formal urban structures. Our argument would be that using public space is not binary and does not of itself divide into formal and informal. The urban landscape is too complex for that. But Bayat (1997) is right to criticise focusing on "civil society" because it ignores the type of social activities that dominate developing countries' urban politics. For him, these activities are quiet encroachment and "a silent, patient, protracted, and pervasive advancement of ordinary people on the propertied and powerful in order to survive hardships and better their lives". This should not be interpreted as Gramsci's (1964) passive revolutionaries but refers to informal need (Devlin, 2017). What drives the activities of disenfranchised groups is not consciously political but their need to survive, and they do not encroach when they use public because public use is the purpose of public spaces. Devlin (2017) and Marx and Kelling (2018) criticise the way studies of informality have been conducted by scholars from the Global North, this is a good way to frame informality, but their argument centres on urban theory's view of informality as a way of understanding "non-western" cities and this is a very tight framing of informality. Equating TA with informality implies no difference between any temporary use of public space and unregulated activity and this is not always so.

Table 1 Activities in the public space related to TA (adapted from Lara H., Melis, & Caputo, 2017)

Category	Economy		Leisure			Sacralisation
Description	Any activity in which a person or group use the public space in order to obtain an economic benefit directly or indirectly		Any activity in which a person or a group use the public space for leisure purposes			Any activity in which a person or a group use the public space for religious purposes
Sub-category	Work	Trade	Sports-games	Artistic expressions	Rest	
Individual or collective	Advertising or promoting services, waiting, engaging or attracting possible clients	Selling or buying products (food, handicraft, clothes, etc.)	Skateboarding, soccer, cards, marbles, hopscotch	Playing music, dancing, painting, acrobatics, reciting and singing	Eating, resting	Processions, praying, lighting candles and putting flowers

A more significant term to conceptualise the relationship between the two concepts is hybridisation, a term provided by the anthropologist Garcia-Canclini (2005). He defines hybridization as "the socio-cultural processes in which discrete structures or practices, previously existing in separate form, are combined to generate new structures, objects, and practices" (2005, p. XXV). For this reason. Garcia Canclini distinguishes hybridisation from fusion and other seeming synonyms; he sees *Mestizaje,*[1] syncretism and fusion as all different from hybridisation. When rural migrants adapt what they know to fit city life, their strategies of reconversion are examples of hybridisation (and not hybridity). By the same token, when craftsmen adapt traditional methods to attract urban buyers, when employers change corporate culture to fit in with new technology, they do the same. And so, do indigenous movements when they change their demands to a form communicable using modern broadcasting and Internet techniques. Simmel (1969) described marginal personalities as cultural hybridisation manifestations involving life in transition between two cultures when one is a full member of neither. These all fit Lara-Hernandez and Melis's (2018) description of TA in urban landscapes. These activities appear more or less harmoniously but pursued each for their own sake produce culture (Eliot, 1949). Lara-Hernandez, Melis and Lehmann (2019) argue that TA is an emerging assemblage product of other assemblages such as the cultural, the legal and the environmental. Deleuze and Parnet (1977) define an assemblage as a "multiplicity which is made up of many heterogeneous terms and which establishes liaisons, relations between them, across ages, sexes and reigns—different natures". Following this argument, we could infer that TA as an emerging assemblage is similar to the process of hybridisation explained earlier.

Taking this idea of temporary association as an assemblage, we returned to the literature to investigate further how urban informality is constructed. We identified two main approaches in the studies focussed on urban informality: the technocratic and the organised bottom-up (see Table 2). This system of classification is very simple and allows us to highlight the main characteristics and flaws of each category. The first approach is more common and holds the view of informality following the rules of the legal, economic or planning system. De Certeau (1984) and Jacobs (1961) are the main critics of this approach by highlighting the importance of the social and of informal everyday life in cities. However, it is not clear what sort of city Everyday Urbanism endeavours to produce, only that it should be constructed in a bottom-up, citizen-driven manner (Devlin, 2017). The second approach encompasses tactical urbanism, DIY Urbanism and the group of studies on community gardens in cities in the USA, Canada and Europe. The Lefebvrian inspired scholars do not see freedom and self-determination as ends unto themselves, rather these concepts are conceived as fuel for a specific revolutionary political project—one that reclaims the means of producing the urban from an oppressive nexus of state and capital (Purcell, 2016).

[1]The term is defined as: (1) Interbreeding and cultural intermixing of Spanish and American Indian people (originally in Mexico, and subsequently also in other parts of Latin America); miscegenation, racial and cultural intermixing. (2) The action or process of such racial and cultural intermixing. rare. (3) Latin American people of mixed American Spanish (especially Mexican) and American Indian parentage, considered collectively, rare (Oxford University Press, 2015).

Table 2 Categorisation of urban informality research approaches

Informality as:	Main advocates	Characteristics	Flaws
Technocratic approach	Crawford (2008), ILO (1972), McFarlane and Waibel (2012) and UN-Habitat (2009, 2013)	It defines informality on the basis of the nature of property rights, planning, infrastructure and level of services. It emerges from the practices of categorisation by the state and/or planning derived by the neoliberalism. It is associated with legal pluralism which holds that plurality of legal systems co-exist, such as state law, religious law, indigenous law, customary laws and local conventions	This approach provides less usefulness to analyse social processes usually context-specific that identifies the so called "urban informality". It evidences the blind spot of planners and designers who were trying to tackle urban issues from a top-down approach by missing the everyday activities and the organic social landscape of the streets. It fails to analyse and describe the informal city because in the literature, both examples are usually conceptualise together while recreation and survival acts belong to a different human nature
Organised bottom-up	Blomley (2004), Burton and Mitchell (2006), Davis (1990), Florida (2012), Low (2010), Low and Smith (2013), Sorkin (1992), Lydon et al. (2014), Lydon & Garcia (2015) and Hou (2010)	It promotes new and creative ways in which urban users through informal tactics can spur and foster urban innovation by increasing the competitiveness of cities that contributes to economic development. Community gardens are part of a broader body of work that attributes political claims to certain kinds of informal use of abandoned or marginal spaces in the industrial city. Such informal uses vary from fulfilling basic needs like housing to a more leisurely activities like artistic activities and all of them share a Lefebvrian conceptualisation of the politics of spatial production	The focus on process (theoretically rigorous though it may be), combined with a notion of desirable outcomes that is only roughly sketched, leaves the door open for practitioners to put these ideas to work in a way that lacks any sort of radical edge or concern for the well-being of disadvantaged communities

Studies of Global North cities do not sufficiently differentiate between formal and informal activities (Chase & Crawford, 2008; Mukhija & Loutaku-Sideris, 2014) because of northern academics' tendency to use a less holistic approach than Global South researchers who view informality as being possessed of a distinctive internal political and spatial logic (McFarlane, 2011; Roy & Alsayyad, 2004a, b; Yiftachel & Hedgcock, 1993). Southern scholars have made clear that informal practices belong to the elite as well as the urban poor. We agree with Devlin, who argues that northern studies of informality lack theoretical tools capable of holistically conceptualising informality. None of Devlin's examples occur ephemerally and spontaneously as TA does, and nor do they take place in public spaces.

The conceptualisation of TA enables us to understand the system of activities within the urban dynamics of a local context in a more holistic manner. "A system of activities occur within a specific system of settings, what does or does not happen in some settings influences what happen or does not happen in other" (Rapoport, 2005: p. 23). There is here a close link to rules about what permitted and forbidden behaviour which form part of culture by dictating who is allowed to do what, where and when they can do it, and why. We agree with Rapoport's (2005) assertion that a system of settings is part of a larger system of settings (street, block, borough and so on).

Discussion and Conclusion

An initial objective of this chapter was to illustrate the nuanced relationship between two complementary concepts, TA and the informal use of public spaces. The first section has discussed the lack of consensus about the definition of informality and how it varies across languages. We have explained that there has to date been a narrow understanding of informality within the academic arena. The second section established the lack of consistency in regard to the term urban informality, and to its study. Therefore, northern approaches to the study of informality are limited in perspective, possibly because of how language forms thoughts and culture and how we interpret environment (Ahearn, 2012). TA has been argued as emerging from local contexts out of the urban environment's culturally driven relationships and dynamics. In addition, the manuscript has provided a categorisation of TA by providing examples from different cities and cultures.

A review is provided of northern approaches to informality and divides them in two main groups. Even when a study describes informal activities in public spaces as "informal behaviour" they provide no evidence. TA can therefore be described as usually misinterpreted as informal practices, while activities, practices or/relations could emerge in any local context. This confirms our earlier point that urban informality studies need a holistic approach. Urban informality study is cross-disciplinary and multidimensional and correlates strongly with other aspects of city and society. The multidimensional concept is complex enough to allow development of research specific to cases and places, suggesting that each study of urban informality should relate only to its specific context. Several authors (Bayat, 1997; Canclini, 2018; Devlin, 2017; García Canclini, 1995; Holston, 2009; Marx and Kelling, 2018) have suggested that the study of urban informality will benefit from multidisciplinary approaches such as the one presented here. However, further studies need to be carried out in this field of research. We agree with researchers such as Clark & Dickson (2003) and Kates et al. (2001) that there is a need to further push the importance of studies about human transformation as being essential to the scientific empirical agenda for the common interest.

Our findings have significant implications for the understanding of informal processes and activities in cities. This work contributes to existing knowledge on urban informality by providing the TA as a medium through which to study the urban landscape at the level of public spaces. Within the field of architecture (landscape, urban design and planning), more information on the design features of the built environment related to TA in any determinate context would help us to establish a greater degree of accuracy in this matter. Some of the questions raised by this study are where does informality intrude, or prevent TA? and how does TA happen without informality being present? There is, therefore, a definite need for a deeper understanding of TA and its relationship with the urban informality of any socio-cultural landscape.

References

Académie Française. (2018). *Académie Française*. Available from: http://www.academie-francaise.fr/le-dictionnaire/la-9e-edition. Accessed November 26, 2018.

Ahearn, L. M. (2012). *Living language: An introduction to linguistic anthropology*. UK, Oxford: Blackwell Publishers Ltd.

Ardura Urquiaga, Á. (2014). Madrid. Espacio público confiscado. La privatización y resignificación del espacio público en los procesos de transformación material de las plazas del centro de Madrid. *Scripta Nova. Revista Electrónica de Geografia y Ciencas Sociales*. [Online] Available from: http://www.ub.edu/geocrit/coloquio2014/AlvaroArduraUrquiaga.pdf.

Auerbach, A. M., LeBas, A., Post, A. E. & Weitz-Shapiro, R. (2018). State, society, and informality in cities of the global south. *Studies in Comparative International Development*. 1–20. [Online] Available from: http://dx.doi.org/10.1007/s12116-018-9269-y.

Bayat, A. (1997). Un-civil society: The politics of the 'informal people'. *Third World Quarterly*. *18*(1), 53–72. [Online] Available from: https://doi.org/10.1080/01436599715055.

Binelli, C. (2016). Wage inequality and informality: evidence from Mexico. *IZA Journal of Labor and Development*. *5*(1), 1–18. [Online] Available from: https://doi.org/10.1186/s40175-016-0050-1.

Blomley, N. (2004). *Unsettling the city: Urban land and the politics of property*. London: UK, Routledge London.

Boer, R., Ramirez-Lovering, D., Moore, T., Heidenreich, S., et al. (2017). The end of informality. *Volume 52*.

Bogue, R. (1991). *Deleuze and Guattari*. New York: Selwood Printing Ltd. West Sussex. [Online] Available from: https://doi.org/10.2307/3684887.

Burton, E. & Mitchell, L. (2006). *Inclusive urban design: Streets for life*. Oxford, UK: Architectural Press. [Online] Available from: http://onlinelibrary.wiley.com/doi/10.1002/cbdv.200490137/abstract%5Cnhttp://issuu.com/filipesilva/docs/ililbykhz2.

Caldeira, T. P. R. (1996). Building up walls: The new pattern of spatial segregation in Sao Paulo. *International Social Science Journal, 48*, 55–66.
Canclini, N. G. (2018). A culture of informality. *Urban studies*, 1–6. [Online] (March 2017). Available from: https://doi.org/10.1177/0042098018782635.
Chase, J. L., & Crawford, M. (2008). *Everyday urbanism*. J. KaliskI (Ed.).
Clark, W. C. & Dickson, N. M. N. (2003). Sustainability science: the emerging research program. *Proceedings of the National Academy of Sciences of the United States of America. 100*(14), 8059–8061. [Online] Available from: https://doi.org/10.1073/pnas.1231333100.
Cordera, R., Ramirez Kuri, P., & Zlccardl, A. (2008). *Pobreza, desigualdad y exclusion social en la ciudad del siglo XXI*. DF, Mexico: Universidad Nacional Autonoma de Mexico.
Crawford, M. (2008). Introduction. In: J. Chase, M. Crawford, & J. Kaliski (Eds.), *Everyday urbanism*. Monacelli Press.
Crossa, V. (2009). Resisting the entrepreneurial city: Street vendors' struggle in Mexico city's historic center. *International Journal of Urban and Regional Research. 33*(1), 43–63. [Online] Available from: https://doi.org/10.1111/j.1468-2427.2008.00823.x.
Crossa, V. (2016). Searching for order on the 'dirty' streets of the city. Displacement of street vendors in colonial and postcolonial Mexico City. In: *Contested Cities*. 2016 Madrid, España. p.
Davis, M. (1990). *City of quartz: Excavating the future in Los Angeles*. London, UK: Verso.
De Certeau, M. (1984). *The practice of every day life*. Steve Rendall (ed.). Los Angeles, USA: University of California Press.
de la Torre, M. I. (2015). Espacio público y colectivo social. *Nova Scientia, 14*(7), 495–510.
Deleuze, G., & Guattari, F. (1989). A thousand plateaus: Capitalism and schizophrenia. In: Massumi, B (ed.). Minneapolis, USA: University of Minnesota Press. [Online] Available from: https://doi.org/10.2307/203963.
Deleuze, G., & Parnet, C. (1977). Deleuze and claire parnet. *Dialogues*. Paris: Flammarion.
Devlin, R. T. (2017). Asking 'Third World questions' of First World informality: Using Southern theory to parse needs from desires in an analysis of informal urbanism of the global North. *Planning Theory,* 1–20. [Online] Available from: https://doi.org/10.1177/1473095217737347.
Díaz Larrañaga, N., Grassi, L., & Mainini, C. (2011). Socialidad: Los modos de apropiación del espacio público. *Question, 1*(29),12. [Online] Available from: http://perio.unlp.edu.ar/ojs/index.php/question/article/view/572/849.
Díaz, F., & Ortiz, A. (2003). *Ciudad e Inmigración: Uso y Apropiación del Espacio Público en Barcelona,* pp. 399–407. [Online] Available from: http://www.insumisos.com/lecturasinsumisas/Apropiaciondelespaciopublico.pdf.
Drummond, L. B. W. (2000). Street scenes: Practices of public and private space in Urban Vietnam. *Urban Studies, 37*(12), 2377–2391. [Online] Available from: https://doi.org/10.1080/00420980020002850.
Duden. (2018). Duden in-for-mell rechtschreibung, bedeutung, definition, synonyme, herkunft, 2018. [Online] Available from: https://www.duden.de/rechtschreibung/informell_unfoermlich_formlos. Accessed November 26, 2018.
Duhau, E., & Giglia, A. (2007). Nuevas centralidades y practicas de consumo en la Ciudad de Mexico: Del microcomercio al hipermercado. *Eure*. [Online] *33*(98), 77–95. Available from: https://doi.org/10.4067/s0250-71612007000100005.
Eliot, T. S. (1949). *Notes towards the definition of culture*. [Online]. Available from: https://books.google.co.uk/books?id=QYR9FNyKMmwC.
Florida, R. (2012). The rise of the creative class. *The Washington Monthley* May, 15–25. [Online] Available from: https://www.os3.nl/_media/2011-2012/richard_florida_-_the_rise_of_the_creative_class.pdf.
Fonseca Rodriguez, J. M. (2015). La importancia y la apropiación de los espacios públicos en las ciudades. *Revista de Tecnologia y Sociedad., 4*(7), 1–11.
García Canclini, N. (1995). Mexico: Cultural globalization in a disintegrating city. *American Ethnologist, 22*(4), 743–755. [Online]. Available from: https://doi.org/10.1525/ae.1995.22.4.02a00050.
Garcia-Canclini, N. (2005). *Hybrid cultures*. Minneapolis, USA: University of Minnesota Press.

Georgieva, K. (2018). The biggest challenge for the future of work hasn' t changed for decades. *World Economic Forum*. [Online] Available from: https://www.weforum.org/agenda/2018/07/technology-can-help-spring-workers-from-the-informality-trap/?utm_source=sfmc&utm_medium=email&utm_campaign=2653547_Agenda_weekly-27July2018-staff&utm_term=&emailType=Newsletter. Accessed July 28, 2018.

Gramsci, A. (1964). Note sul Machiavelli, sulla politica e sullo stato moderno. G. Einaudi.

Graumann, C. F. (1976). The concept of appropriation (aneignung) and modes of appropriation of space. *6*(2), 301–313. [Online] Available from: http://iaps.scix.net/cgi-bin/works/Show?iaps_00_1976_009.

Henle, P. (Ed.). (1965). Language, thought, & culture. University of Michigan Press.

Hernandez Mendo, A., & Morales Sanchez, V. (2008). Una revisión teórica: ocio, tiempo libre y animación sociocultural. *Efdeportes.com*. [Online]. (27), 9–12. Available from: http://www.efdeportes.com/efd127/una-revision-teorica-ocio-tiempo-libre-y-animacion-sociocultural.htm.

Holston, J. (2008). *Insurgent citizenship: Disjunctions of democracy and modernity in Brazil*. New Jersey: USA, Princeton University Press.

Holston, J. (2009). Insurgent citizenship in an era of global urban peripheries. *City and Society*, *21*(2), 245–267. [Online]. Available from: https://doi.org/10.1111/j.1548-744X.2009.01024.x.

Hou, J. (2010). *Insurgent public space: guerrilla urbanism and the remaking of contemporary cities*. New York: USA, Routledge.

International Labour Office. (1972). *Employment, incomes and equality*.

Iveson, K. (2013). Cities within the City: Do-It-Yourself Urbanism and the Right to the City. *International Journal of Urban and Regional Research, 37*(3), 941–956. [Online] Available from: https://doi.org/10.1111/1468-2427.12053.

Jacobs, J. (1961). *The Death and Life of Great American Cities*. [Online] Available from: https://doi.org/10.2307/794509.

Kates, R. W., Clark, W. C., Corell, R., Hall, J. M., et al. (2001). Sustainability Science. *Science's Compass, 292*(April), 641–642.

Kim, A. M. (2013). *Sidewalk city*. Chicago, US: The University of Chicago Press. [Online] Available from: https://doi.org/10.7208/chicago/9780226119366.001.0001.

Kinder, K. (2014). Guerrilla-style defensive architecture in detroit: A self-provisioned security strategy in a neoliberal space of disinvestment. *International Journal of Urban and Regional Research, 38*(5), 1767–1784. [Online] Available from: https://doi.org/10.1111/1468-2427.12158.

Kuppinger, P. (2019). Informal place-making: Mosques, muslims, and Urban innovation in Germany. *The palgrave handbook of bottom-up Urbanism* (pp. 149–162). Cham: Palgrave Macmillan.

Lara-Hernandez, J. A. & Melis, A. (2018). Understanding the temporary appropriation in relationship to social sustainability. *Sustainable Cities and Society, 39C*, 1–14. [Online] Available from: https://doi.org/10.1016/j.scs.2018.03.004.

Lara-Hernandez, J. A., Melis, A. & Caputo, S. (2017). Understanding spatial configuration and temporary appropriation of the street in latin American cities: The case of Mexico City Centre Hocine Boudagh, Antonella Versaci, Ferdinando Trapani, Migliore, M. et al. (Eds.), *Advances in science, technology & innovation* (pp. 153–173).

Lara-Hernandez, J. A., Melis, A. & Coulter, C. M. (2018). Using the street in Mexico City centre: Temporary appropriation of public space versus legislation governing street use. *Journal of Public Space, 3*(3), 25–48. [Online] Available from: https://doi.org/10.32891/jps.v3i3.1135.

Lara-Hernandez, J. A., Melis, A. & Lehmann, S. (2019). Temporary appropriation of public space as an emergence assemblage for the future urban landscape: The case of Mexico City. *Future Cities and Environment, 5*(1), 1–22. [Online] Available from: https://doi.org/10.5334/fce.53.

Leary-Owhin, M. (2015). A fresh look at lefebvre's spatial triad and differential space: A central place in planning theory? In: *2nd Planning Theory Conference University of the West of England*, pp. 1–8 (2015).

Lefebvre, H. (1971). De lo rural a lo urbano. In Gonzalez-Pueyo, J (Ed.), Barcelona, España: Ediciones Peninsula.

Lefebvre, H. (1992). *The production of the space* (1996th ed.). Massachusetts, Cambridge: Blackwell Publishers Ltd.
Lehmann, S. (2015) Low carbon cities: More than just Buildings. In: Lehmann, S. (Ed.), *Low carbon cities: Transforming Urban systems* (pp. 1–56).
Loukaitou-Sideris, A., & Ehrenfeucht, R. R. (2009). *Sidewalks conflict and negotiation over public space*. USA: Massachusetts Institute of Technology.
Low, S. M. (2010). *On the plaza: The politics of public space and culture*. Texas, USA: University of Texas Press.
Low, S., & Smith, N. (2013). *The politics of public space*. New York, USA: Routledge.
Lucia Bayer, M. F. (2016) Conlictos en torno al espacio urbano y dinámicas de apropiación en el Río de Janeiro Olímpico. El caso de Villa Autódromo. *Congreso Internacional Contested*.
Lutzoni, L. (2016). In-formalised urban space design. Rethinking the relationship between formal and informal. *City, Territory and Architecture, 3*(1), 20. [Online] Available from: https://doi.org/10.1186/s40410-016-0046-9.
Lydon, M., Bartman, D., Garcia, T., Preston, R., et al. (2014). *Tactical urbanism*. New York, US: Street Plans.
Lydon, M., Bartman, D., Woudstra, R. & Khawarzad, A. (2012). *Tactical Urbanism: Short Term Action| Long Term Change*. beta. New York, US: Street Plans Collaborative.
Lydon, M., & Garcia, A. (2015). A tactical urbanism how-to. In: *Tactical urbanism*. Springer. pp. 171–208.
MacDonald, R. & Shildrick, T. (2007). Street corner society: leisure careers, youth (Sub)culture and social exclusion. *Leisure Studies, 26*(3), 339–355. [Online] Available from: https://doi.org/10.1080/02614360600834826.
Madanipour, A. (1996). *Design of urban space. An inquiry into socio-spatial process*. New York, US: John Wiley & Sons Inc.
Marx, C. & Kelling, E. (2018). Knowing urban informalities. *Urban Studies. 1*(16). [Online] Available from: https://doi.org/10.1177/0042098018770848.
Mayer, M. (2007). *Contesting the neoliberalization of urban governance* (pp. 90–115). Contesting neoliberalism: Urban frontiers.
McAuslan, P. (2006) Legal pluralism as a policy option: Is it desirable, is it doable? *Land Rights for African Development From Knowledge to Action,* 9–10 (2005). [Online] Available from: http://www.capri.cgiar.org/wp/..%5Cpdf%5Cbrief_land-04.pdf.
McFarlane, C. (2011). The city as assemblage: Dwelling and urban space. *Environment and Planning D: Society and Space, 29*(4), 649–671. [Online] Available from: https://doi.org/10.1068/d4710.
McFarlane, C., & Waibel, M. (2012). *Urban informalities: reflections on the formal and informal*. UK, London: Ashgate Publishing Ltd.
Merriam-Webster. (2018). *Dictionary by Merriam-Webster* (2018). [Online]. Available from: https://www.merriam-webster.com/. Accessed November 26, 2018.
Mitchell, D., & Staeheli, L. A. (2006). Clean and safe? property redevelopment, public space, and homelessness in downtown San Diego. *The politics of public space*, 143–175.
Moatasim, F. (2018) Entitled urbanism: Elite informality and the reimagining of a planned modern city. *Urban Studies,* 1–17. [Online] Available from: https://doi.org/10.1177/0042098018767011.
Monnet, J. (1995) *Usos e imágenes del centro histórico de la Ciudad de México*. Centro de estudios mexicanos y centroamericanos (ed.).
Moudon, A. V. (1991). *Public streets for Public use*. US, New York: Columbia University Press.
Mouffe, C. (2014). Agonística. Pensar el mundo políticamente. Laclau, S. (Ed.), *Mexico, DF, Fondo de Cultura Económica de Argentina*.
Noschis, K., Dosio, M.-J., Feddersen, P., & Triantis, E. (1978). Appropriation of space: a method and two case studies. *Ekistics, 45*(273), 451–466.
Oxford University Press. (2015). *Oxford english dictionary online* (2015). Oxford English Dictionary. [Online] Available from: http://dictionary.oed.com. Accessed May 12, 2015.

Pasquetti, S. & Picker, G. (2017) Urban informality and confinement: Toward a relational framework. *International Sociology, 32*(4), 532–544. [Online] Available from: https://doi.org/10.1177/0268580917701605.

Peattie, L. (1987). An idea in good currency and how it grew: The informal sector. *World Development, 15*(7), 851–860. [Online] Available from: https://doi.org/10.1016/0305-750x(87)90038-6.

Pol Urrútia, E. (2002). El modelo dual de la apropiación del espacio en Mira, R., Sabucedo, JM y Romay, J. *Psicología y Medio Ambiente. Aspectos psicosociales, educativos y metodológicos,* 123–132.

Portal, M. A. (2009). Las creencias en el asfalto. La sacralización como una forma de apropiación del espacio público en la ciudad de México. *Cuadernos de Antropología Social, 30,* 59–75.

Purcell, M. (2002) Excavating Lefebvre: The right to the city and its urban politics of the inhabitant. *GeoJournal, 58,* 99–108. [Online] Available from: https://doi.org/10.1023/b:gejo.0000010829.62237.8f.

Purcell, M. (2016). For democracy: Planning and publics without the state. *Planning Theory,* 15(4), 386–401. [Online] Available from: https://doi.org/10.1177/1473095215620827.

PWN. (2018). *Słownik Języka Polskiego.* 2018. [Online] Available from: https://sjp.pwn.pl/szukaj/nieformalny.html. Accessed November 26, 2018.

Ramirez-Lovering, D. (2008). *Opportunistic urbanism.* Melbourne, Australia: RMIT Press.

Rapoport, A. (2005). *Culture, architecture, and design.* Chicago: Locke Science Publisher.

Real Academia Española. (2017). *Informal|definición de informal.* 2017. [Online] Available from: http://dle.rae.es/?id=LXum6xU. Accessed October 25, 2018.

Rios, M. (2014) Implications for Urban Design. In: Vinit Mukhija & Anastasia Loukaitou-Sideris (eds.). *The Informal American City.* Mit Press. pp. 173–191.

Roy, A. (2005) Urban informality: Toward an epistemology of planning. *Journal of the American Planning Association, 71*(2), 147–158. [Online] Available from: https://doi.org/10.1080/01944360508976689.

Roy, A. (2009) Strangely familiar: Planning and the worlds of insurgence and informality. *Planning Theory, 8*(1), 7–11. [Online] Available from: https://doi.org/10.1177/1473095208099294.

Roy, A. (2012) Urban Informality: The production of space and practice planning. *The Oxford Handbook of Urban Planning,* (July 2018), 1–17. [Online] Available from: https://doi.org/10.1093/oxfordhb/9780195374995.013.0033.

Roy, A., & Alsayyad, N. (2004a). *Urban informality: Transnational perspectives from Midle East. Latin America and South Asia.* Oxford, UK: Lexington Books.

Roy, A., & AlSayyad, N. (Eds.). (2004b). *Urban informality.* Maryland: Lexington Books.

Sarmiento, H. & Tilly, C. (2018). Governance lessons from urban informality. *Politics and Governance, 6*(1), 199. [Online] Available from: https://doi.org/10.17645/pag.v6i1.1169.

Seaman, J. & McLaughlin, S. (2014). The importance of outdoor activity and place attachment to adolescent development in Coös County, New Hampshire. *Building knowledge for families and communities, 208*(37). [Online] Available from: http://scholars.unh.edu/carsey/208/.

Seghezzo, L. (2009). The five dimensions of sustainability. *Environmental Politics, 18*(4), 539–556. [Online] Available from: https://doi.org/10.1080/09644010903063669.

Sen, A. (1999). *Development freedom.* New York, USA: Oxford University Press.

Simmel, G. (1969). Essays classic on the culture of. In: Richard Sennet (Ed.), *Classic essays on the culture of cities.* New Jersey: Prentice-Hall Inc (pp. 47–60).

Simpson, P. (2011). Street performance and the City: Public space, sociality, and intervening in the everyday. *Space and Culture, 14*(4), 415–430. [Online] Available from: https://doi.org/10.1177/1206331211412270.

Sorkin, M. (1992). *Variations on a theme park: The new American city and the end of public space.* Macmillan.

TRECCANI. (2018). *Informale in vocabolario—treccani.* 2018. [Online] Available from: http://www.treccani.it/vocabolario/informale/ Accessed 25 October, 2018.

UN-Habitat. (2009). *Planning Sustainable Cities—Global Report on Human Settlements 2009.* [Online]. Available from: https://doi.org/10.1007/s13398-014-0173-7.2.

Un-Habitat. (2013). *Planning and design for sustainable urban mobility: global report on human settlements 2013*. Available from: [Online] http://www.indiaenvironmentportal.org.in/content/382038/planning-and-design-for-sustainable-urban-mobility-global-report-on-human-settlements-2013-policy-direction.

Van Dale (2018). *Gratis woordenboek*. 2018. [Online] Available from: https://www.vandale.nl/gratis-woordenboek/nederlands/betekenis/informeel#.W_v-L_n7SUl. Accessed November 26, 2018.

Yiftachel, O., & Hedgcock, D. (1993). Urban social sustainability: the planning of an Australian city. *Cities, 10*(2), 139–157.

Yory, M. C. (2011) El Concepto de Topofilia entendido como Teoría del Lugar. *Revista Barrio Taller,* 1–17.

Yory García, C. M. (2003). Topofilia, ciudad, territorio: una estrategia pedagógica de desarrollo urbano participtivo con dimensión sustentable para las grandes metrópolis de América Latina en el contexto de la globalización: 'el caso de la ciudad de Bogotá'. Universidad Complutense de Madrid. [Online] Available from: http://biblioteca.ucm.es/tesis/ghi/ucm-t26725.pdf.

Ongoing Appropriation: Invisible Seattle and Red May

Keith Harris

Abstract During the last few days of April, 2017, I began to see announcements online and around Seattle about a festival called Red May. With speakers drawn from both academia and the arts, the events promised to bring diverse radical thinkers—including Kathi Weeks, Michael Hardt, Jason Read, China Miéville, Geoff Mann, and Nisi Shawl—into conversation in various venues around the city: cafes, art galleries, Town Hall, the Grand Illusion Cinema. As I attended events and met the organizers, I discovered that these events were a mature and explicitly anti-capitalist development of a project from the early 1980s called Invisible Seattle, which drew heavily on Surrealism and Situationism. The playfulness of Invisible Seattle lives on in Red May as a roving festival of ideas and imagination which seeks to make the month "the public home for Left reflection absent since the Occupy Movement was evicted from the heart of America's cities" (Wohlstetter, 2017). This chapter serves as both a history of the movement between the two groups and their work, as well as an application of, and a theoretical contribution to, the concept of "transindividuality" (Read, 2016). The central argument is that although urbanists and activists often focus on the spatial dimensions of occupation, the temporal dimension is perhaps even more crucial for the constitution of collective subjectivity.

Keywords Invisible Seattle · Red May · Temporality

Introduction

Let's start with two experiences that are separated by about 18 months. On the one hand, during the last week of April, 2017, I began to see announcements online and around Seattle about something called Red May. With speakers drawn from both academia and the arts, it promised to bring diverse radical thinkers—including Kathi

K. Harris (✉)
University of Washington, Washington, USA
e-mail: gkh2@uw.edu

A. Melis et al. (eds.), *Temporary Appropriation in Cities*,
https://doi.org/10.1007/978-3-030-32120-8_5

Fig. 1 Joshua Clover (University of California—Davis) presents on the general law of capitalist accumulation from Marx's *Capital* in the Northwest Film Forum

Weeks, Michael Hardt, Jason Read, China Miéville, Geoff Mann, and Nisi Shawl—into conversation in various venues around the city: cafés, art galleries, community space, bookstores, and art-house cinemas (Fig. 1). I came to learn that the month's activities were organized around the Red May Catechism, which begins thus:

> Maybe we can't move beyond capitalism by next week, but we can sure as hell take a vacation from it, from the buzzwords that Seattle is particularly susceptible to: "high tech," "entrepreneurial," "the innovation society," "the gig economy." If everything around us chants "there's no alternative to capitalism"; if all we're told is that history is over and ends with us, don't you think simple mental hygiene demands that we try to think outside of that box for, at the very least, one month a year? (Wohlstetter, 2017)

On the other hand, a year and half later, I read a short *New York Times* article entitled "The Existential Void of the Pop-Up 'Experience.'" The author, Amanda Hess, begins with a beautifully snarky line: "One evening this summer, I left work around 6, ducked into a storefront in Midtown Manhattan and stood at the back of a line, waiting to become myself." (Hess, 2018). She was poised to enter the so-called "Rosé Mansion" as part of her project of exploring the constellation of such pop-up events in New York, which she characterizes as "canny, if cynical" responses to Walter Benjamin's essay "The Work of Art in the Age of Mechanical Reproduction" (1969). Whereas for Benjamin, the reproduction of art diminished the aura of the original, the pop-up event's aura is fueled by the visitors' playful behavior, by their activity.

The cynicism of these events, these "Instagram Museums," is expressed, for Hess, in design elements such as the ball pits that are cribbed from McDonald's PlayPlaces across North America. The products of these events, which are evidently "designed to fit the shrunken-down Instagram grid," are nothing more than another disposable visual blip on a perpetually shape-shifting feed.

This chapter takes these two experiences as poles delineating a spectrum across which temporary appropriations of space might be organized, in an effort to think about what one might traditionally call the process of becoming a collective political subject. However, following recent work on the concept of "transindividuality," notably by the philosopher and political theorist Jason Read, I want to step away from the concept of building a revolutionary political subject and try to escape what he calls, citing Yves Citton (2012), a "'grand politics,' a politics founded on a rupture, a radical event, and a subject that could sustain it" (Read, 2016: p. 288). Instead, Read asserts that politics "has as its condition not the event, not some ruptural break with the existing order, but the *tensions and divisions within the existing order*" (Read, 2016: p. 289, my emphasis). For what is a month-long vacation from capitalism—as Red May describes itself—if not an attempt to augment tensions within an existing order?

I therefore focus primarily on the temporal domain rather than the spatiality of these appropriations and for two reasons. First and practically, as Lara-Hernandez and Melis (2018) note, the temporality of appropriation is an under-researched yet essential dimension of for understanding the contemporary urban experience. Second and theoretically, time is particularly important because it is the domain of *becoming*, in the philosophical sense (cf. Deleuze and Guattari, 1987). To this end, I focus on the Red May end of the spectrum rather than that of the Rosé Mansion. However, I also discuss Invisible Seattle—the predecessor of Red May—as a powerful reminder that temporary appropriations can be both playful and intellectually stimulating, and that, furthermore, they can function beyond the physical space in which they occur: for example, Invisible Seattle's critical and playful interventions also extended into mediatic, virtual, and imaginative space. But even more importantly, I emphasize that the ongoing nature of both the Invisible Seattle and Red May projects is key to magnifying the tensions that everyone experiences in the maelstrom of ever-shifting contemporary urban environments, and ultimately cultivating the gesture of appropriating space for creative and politically-charged endeavors.

Methodologically, my approach is mixed. Theoretically, I rely on close readings of various thinkers such as Gilles Deleuze, and on commentary and elaborations by theorists who have furthered his project, notably Todd May and Read. Empirically, I draw on my own experiences of Red May events, as well as video and audio archives of events that I did not attend, informal conversations with the organizer, Philip Wohlstetter, and various archives of "Invisibilia" related to the Invisible Seattle project, including a novel, press coverage, memoirs, and interviews.

Normative Ontology and the Politics of Time

Perhaps the most striking observation that Todd May (1996) makes in his essay on Deleuze's conception of time actually has nothing at all to do with time, but ontology. He writes:

> For Deleuze, ontology is a normative endeavor. The point in doing ontology is not to arrive at an understanding of the way the world is structured, but to be able to take up a certain viewpoint toward the world in order to engage in certain ways of living (May, 1996: p. 294)

This approach is crucial for how I want to think about a specific flavor of Temporary Appropriation (TA) that I call Ongoing Appropriation (OA). I am not particularly invested in terminology here, nor am I concerned with the design of urban spaces or particular sites of appropriation, but I am very interested in establishing new political alliances, or what Read would call new "transindividuations," through these appropriations—particularly in events that occur over an extended period of time. This is why I find Red May so fascinating: it hosts events in any available space, without getting caught up with the particulars of the space itself; it makes itself at home in art-house cinemas as well as in posh bars adjoining high-tech co-working spaces (Fig. 2). Therefore, the impetus behind trying to define OA is to think about the political problem of building political alliances in any-space-whatsoever. This problem is, by definition, a temporal problem in Deleuzean philosophy, where the process of becoming occurs in time rather than in space.

Fig. 2 Sianne Ngai (University of Chicago, second from left) and Chris O'Kane (The City University of New York, far left) discuss commodity fetishism in The Cloud Room

It is helpful to briefly return to Deleuze's (1994a) reading of Henri Bergson—his primary writing on time—to illustrate the fundamental difference between space and time, based on the two kinds of multiplicities that Bergson defines. The first type belongs to space and "it is a multiplicity of exteriority, of simultaneity, of juxtaposition, of order, of quantitative differentiation, of *difference in degree*" (1994: p. 32). These types of differences are *actual*—in the sense of Deleuze's metaphysics, which divides the real into the "actual" and the "virtual," rather than the more common division that is made between the "real" and the possible" (cf. Deleuze, 1994a, b); such differences can be perceived and one can judge between them using comparison (smaller/larger, lighter/darker, heavier/lighter, faster/slower, etc.). The second type of multiplicity, however:

> appears in pure duration: It is an internal multiplicity of succession, of fusion, of organization, of heterogeneity, of qualitative discrimination, or of difference in kind; it is a virtual and continuous multiplicity that cannot be reduced to numbers (Ibid.).

The difference (in kind) between the actual and the virtual is the same as the difference (in kind) between space and time. For example, I am writing this looking out a window of Café Solstice on Capitol Hill in Seattle, Washington, on December 15, 2018. I see a man and a woman walking toward one another on the sidewalk in front of a community garden. I can see actual differences—he is taller and has a longer stride; her hair is longer and her skin is darker—but virtual differences, differences in their *capacities* remain invisible to me: she could be a talented pianist and he a mediocre poet, but she might be teetering on exhaustion with her instrument while his abilities may continue to flourish, or vice versa. *Only time will tell.*

Memory also illustrates the difference between the actual and the virtual, and helps establish Deleuze's commitment to Bergson's monistic vision of time. Using the example of the two pedestrians: given that I just perceived them, I can access the memory in what Deleuze calls psychological memory, or memory that is present and actual. However, if I reread this chapter in six months, these words will cause me to plunge into the virtual, or what Deleuze calls ontological memory—that is, the singular, non-individual past that we can all access. This ontological memory is the virtual and this is, for Bergson, how all memory works. *Where were you on September 11, 2001?* There was an attack on the World Trade Center in New York City and I found out about it sitting next to Andy Berman in a civil engineering course at Texas A&M University. All particular memories are different, but the fact that two planes were flown into the towers and that I was sitting in College Station, Texas, next to another student is part of the past, and he, Andy, could dive into the virtual, and actualize his memories of the same morning. Whereas the actual differences (in degree) that I perceived between the two pedestrians played out in space, the differences (in kind) between, for example, how my understanding of these ideas has unfolded since seeing the pedestrians, or how my and Andy's capacities for engineering or thinking about Deleuze have developed since sitting in that classroom 17 years ago, are instantiations of difference-in-itself.

But back to OA: if the normative position of this concept is that the goal is to facilitate a revolutionary process of transindividuation—that is, the simultaneous emergence of both characters and a collective capable of, for example, enacting something like a general strike[1]—then it is in this unfolding of time, of duration, where such a process occurs. I now briefly introduce Read's conception of transindividuality before discussing Red May and its predecessor, Invisible Seattle, in more detail.

The Politics of Transindividuality

Read's central concern is challenging what he calls the false binary between the notions of the individual and the collective. He begins by describing the present as an era in which the individual takes precedence over the collective: in politics, this is expressed in an ethic of individual rights and freedoms; in economics, maximizing one's own utility is of primary importance; and in the cultural realm, the "ideal of complete and utter self-expression and independence" (Read, 2016: p. 1) rules. Yet a contradiction arises when this focus on individuality is confronted with the reality of the abstract and impersonal forces of the contemporary world, such as technology, globalization, and the media. It is precisely the tension between "an ideal of the individual and a dim awareness of social relations, of social conditions, that can only be perceived as a hostile or at the very least inevitable force" (Read, 2016: p. 2) that constitutes the dominant contemporary mindset. Read then embarks in a bi-directional reading of theorists who he sees as challenging the very binary, starting—in true Deleuzoguattarian fashion—in the middle, with the work of Gilbert Simondon.

Simondon forms a pivot for Read's interrogation of transindividuality for two reasons. First, rather than taking the individual as the starting point, Simondon explicitly posits individuation as process in which a metastable pre-individual state is resolved into an increasingly stable, yet still *temporary*, individual—a phenomenon that he attributes to all forms of individuation (biological, physical, psychic, etc.). Second, the individuation of a particular character does not occur independent of a group, nor does the individuation of a group occur independent of that of individuals; instead, the two processes are co-constitutive and utilize the same set of metastable ingredients (affects, emotions, and perceptions). Read's project unfolds by tracking theorizations of transindividuality in both pre- and post-Simondonian thinkers (Spinoza, Hegel, and Marx in the former category, and Étienne Balibar, Bernard Stiegler, Paolo Virno, and Maurizio Lazzarato in the latter). The details of this work are obviously outside the purview of this chapter, but I want to briefly focus on some of his concluding

[1]During a recent conversation, Wohlstetter hypothesized that the 1999 WTO protests were the second general strike in Seattle, with the first being in 1919. His question was: what would a third general strike look like?

thoughts to frame the transindividuation that Red May and Invisible Seattle cultivate through their extended durations.

Read's rendering of Stiegler's and Virno's accounts of the relationship between transindividuation and contemporary capitalism is a powerful starting point for thinking about my cases. Stiegler, for example, is a key reference here because he "stresses the essentially temporal character of every transindividual individuation" by highlighting the fact that it "is a process, a movement from an initial metastable condition, defined by tensions and relations [among affects, emotions, and perceptions], towards a future" (Ibid., 169). Yet he, Stiegler, focuses on the consumption side of the modern economy, paying particular attention to contemporary consumer culture's efforts to coordinate our affects, emotions, and perceptions in ways that diminish the complex individuation process that produces knowledgeable citizens. Read is quick to point out that, while consumption is indeed an important vantage point, it is not the entirety of the economy. For this reason, his exploration of Virno's writings on transindividuality and the production process—particularly the "constitution of a new collectivity and a new individuality" (Ibid., 214) that is made possible in the post-Fordist era of increasingly cooperated production—is illuminating. But Read is well aware that the economy does not have a monopoly on the individuation process, and notes that these forces constantly interact with those in the traditional political realm. It is here, in his discussion of Citton's work, that I see a both a positive political task that connects directly to Red May and Invisible Seattle. He writes:

> Citton's real interest in the concept [of metastability] lies in how it makes it possible to grasp not the constitution of economic and political relationships, but their transformation. Thinking gestures, affects, and myths as different transindividual relations makes it possible to see the way in which one aspect not only can be in constitutive tension with the others, but also can break free, constituting the basis for a new metastable relation. A gesture can be reproduced, duplicated, and communicated, and the boundaries of this communication often exceed the boundaries drawn by myths of community, and the affects that structure a given individuation (Read, 2016: p. 279).

This notion of gesture is, I think, the key thing to be learned from Red May and Invisible Seattle. The operative gesture? *Appropriating any-space-whatsoever toward specific creative ends, be they explicitly political or "merely" artistic.* The extended durations of these occupations are crucial because they give participants multiple opportunities to enact this new gesture, to build up new habits, and to increase the probability of making meaningful and lasting alliances with others, even in an epoch that feels mired in what Mark Fisher (2009) has called "capitalist realism." In other words, they create the conditions for a new process of transindividuation. The next two sections explore some of the ways that these ongoing forms of appropriation have taken and continue to take place—and take time, so to speak—in Seattle.

Red May

In his opening comments to Red May 2017, Philip Wohlstetter says:

> Welcome to Red May, your one-month vacation from capitalism. We have 42 events in 30 days…[Y]ou [can] think of it is a floating conversation that moves in and out of kind of diverse spaces like a movie theater, a bar, a kind of elegant terrace…by the end of the day everybody is pretty blitzed, but Marx has kind of sunk down into your under-consciousness, never to be dislodged, even by intense, enhanced interrogation (Swenson, 2017).

It is the sustained engagement of these events and the lodging of the ideas in the participants that makes these events so compelling. As Wohlstetter told me in conversation, the idea is more along the lines of a month-long teach-in. And in response to those who have suggested that it would be more manageable to organize events over a period of three or four days days, he remarked, "Three days? That's a conference!" In 2017, these 42 events took place in independent cinemas, bars and restaurants, bookstores, art galleries, and community spaces around Seattle.

The first event I attended—"Red Planets: The Left Turn in Science Fiction"—took place in The Grand Illusion Cinema, a 70-seat art-house theater in the University District. It put the science fiction author Nishi Shawl, critical theorists Jason Read, and cultural studies scholar Tom Foster in conversation to discuss "extrapolation, cognitive estrangement, the ability to unearth the fragment of the future in the present or the past alternative that the capitalist present killed" (Red May, 2017). As relatively new reader of science fiction at the time, I was intrigued by this perspective and left with the name Kim Stanley Robinson cemented in my brain—an author who has since reframed how I imagine the intersection of politics, technology, and design. His Mars Trilogy, which explores the colonization of Mars, the fractured political commitments among various groups of colonists, and the efforts to terraform the planet was a common point of reference for all the panelists on how the genre of science fiction can address the intersections between radical politics, technology, and social organization.[2]

Another event that I attended took place in the Vermillion art gallery, in Capitol Hill, and it brought together the political theorists Kathi Weeks and Michael Hardt for a panel entitled "Should We Love Our Work?" Approximately 50 participants sat at tables and stools surrounding the speakers, and given the theme and timing of the event—during a doctoral seminar I was teaching on ethics and during a week in which we were discussing the ethics of teaching and the politics of non-tenured academic labor—I brought my class to the event. The combination of exposure to radical feminist political theory and various interpretations and questions from people precariously employed in the arts and academia—undoubtedly fueled in part by the charming environment and the libations from a fully-stocked bar—made for a reprieve from the sometimes stifling atmosphere and relative distance from the "real-world" that characterizes academia. For me, it fell somewhere between group

---

[2]McKenzie Wark addresses Robinson's work explicitly in *Molecular Red: Theory for the Anthropocene* (2017).

therapy, political scheming, and socializing with friends from different walks of life who do not typically spend time together.

In 2018, events again took place in a wide range of venues, and occurred in performance/community spaces or art galleries, but also in bookstores, university classrooms and lecture halls, and even a church. Asad Haider, a soon-to-be graduate from the University of California—Santa Cruz's History of Consciousness doctoral program, read portions of his critique of the contemporary, individualistic view of identity politics to an audience in the basement of the Elliott Bay Book Company, citing major thinkers like Michel Foucault and Judith Butler, alongside lesser known but influential groups such as the Combahee River Collective, a black lesbian militant group which was formed in Boston in 1974 (Haider, 2018). A few days later, Nikhil Pal Singh, Professor of Social and Cultural Analysis and History at NYU, shared excerpts from his recent book, *Race and America's Long War* (2017), which interrogates the way that policing and war produce the idea of race. For example, drawing on colonial law in Virginia, he described how members of indigenous tribes and blacks were stripped bare during punishment by whipping, while whites were permitted to retain some dignity by remaining clothed. Similar empirical details, such as first-hand accounts from mid-19th century slave patrollers in Georgia, fueled his argument that the rationalization of such crimes us a crucial moment in the birth of the United States. Later in the month, Vancouver activists and scholars Matt Hern and Am Johal presented portions of their book—which also includes illustrations by the well-known journalist and cartoonist, Joe Sacco—on the ecological effects of oil extraction from the tar sands in northern Alberta from this same stage. The diversity and depth of such events, with these authors presenting work published by Verso, University of California Press, and MIT Press, respectively, delivered graduate seminar-quality content and discussion to the city's premier independent bookstore, located steps from fashionable restaurants, concert venues, and increasingly luxurious housing for the growing ranks of tech workers.

The ongoing nature of these events provides multiple opportunities for encounters with the material being discussed and other participants, as well as for the shared experience of affective modulations and the exchange of rational ideas that are part and parcel of the parallel individuation of both individuals and collectivities. From a purely pragmatic angle, the quantity of events and their location outside of a university context and in basement of a charming bookstore, adorned with bookshelves, a small stage and podium, and the large red flags that the Red May organizers carry between events, both increases the probability for productive encounters, and establishes many opportunities to enact the gesture of gathering and serious engagement that, as Stiegler stresses, is less common in our contemporary consumer culture. Moreover, while the spaces are indeed important elements, their role as containers for such events outweighs any of their formal properties. Instead, the diversity of spaces might be better interpreted as an invitation to think of any space as being ripe for appropriation.

In sum, the Red May events offer participants a variety of ways of interacting with the cutting edges of contemporary critical theory in diverse locations around Seattle. From the perspective of OA, the extended duration of these events increases the likelihood that participants can form meaningful and potentially politically active alliances. In contradistinction to the banal and consumer-friendly pop-up events that Amanda Hess documents in her article for the *Times*, the events of Red May are designed to spur critical thinking and collective consciousness. Wohlstetter's explicit political focus in his work on Red May is, however, a transformation of some of his earlier–though no less creative and public–work with Invisible Seattle, to which I now turn.

Invisible Seattle

August 2, 1983, noon, in the Seafirst Bank Plaza, across from Seattle Public Library (not the Rem Koolhaas building that opened in 2004, but the International-Style building that replaced the 1904 Carnegie library in 1960): Ladders, construction cones, surveyor's tape, wheelbarrows, buckets and lumber occupy the sterile corporate environment. Workers mill about in hardhats and white jumpsuits, stenciled with words: SKYSCRAPERS, DONUTS, BOREDOM. Upon hearing the foreman's whistle, they carry buckets of vowels and push wheelbarrows of consonants to their workstations. The Bullhorn Crew harangues passersby and explains the event; interviewers collect words from participants according to precise rules regarding posture and body movements. The foreman then distributes words to workers, who build them by hanging them on lumber spanning between ladders, yelling out things like "hit me again," upon completion of each (Fig. 3). If someone needs a letter that is not in their bucket or wheelbarrow, the script advises them to "call out for it and someone will throw it to you like a rivet" (Invisible Seattle, 1987: p. 225). The script also informs the workers that they should continue until the whistle blows for a break; everyone is to take out their beer when someone shouts "Hey, it's Schiller time!" before heading back to their stations.

This event marked the beginning of the collective authorship of *Invisible Seattle: The Novel of Seattle, by Seattle*. In a press release the previous day, the arts group Invisible Seattle announced that they were "offering the people of Seattle the chance to become the author of their own book, their own city, their own destiny" (Invisible Seattle 1987, frontispiece). They would spend the next month collecting data in cafés, bookstores, libraries, street corners, and the Kingdome, which would be loaded into a special database on their literary computer, Scheherezade II. Subsequently, during the Bumbershoot Arts festival, from September 2–5, workers—as well as the public at times—would assemble information from the database into the agreed upon format for the first collective novel of a city, by a city. A detective novel of sorts, one in which people and places disappear, including Proteus, the agent that the reader is tracking. "We write novels the way they used to build cathedrals," says Invisible Seattle.

The information collected throughout the month of August produced the setting of the story. For example, a literary worker might stop a pedestrian an ask, "What is disappearing in Seattle?" Answers clearly referred to some of the smaller, eccentric restaurants, such as The Doghouse, an all-night restaurant, which features heavily in

Fig. 3 Literary workers assemble characters' names during the kickoff event for Invisible Seattle's collectively authored novel

the book but vanishes toward the end (it really closed in 1994 and its former building was recently demolished to make space for Amazon's expanding corporate campus). Responses to questions such as, "What were you doing at 10:43 pm?" appear across the novel:

> The city sits in the bus station with eleven minutes to wait for the arrival from Ellensburg, eating ice cream, leaning on its elbows three quarters of the way down the bar, thinking about lounging around in the living room, playing pool or playing with its spouse, watching the Playboy Channel, discussing with its roommate, Lee, about how to produce Ultra Violet Cabaret, and worrying about collecting its bags from the United flight that arrive from Minneapolis (via Denver) at 10:30. (Invisible Seattle, 1987: p. 101)

Descriptions of citizens' cars populate a drive-in movie theater, where you—the reader, the detective—are trying to make love to Terry, one of your informants: A "1971 Pontiac GTO…a lemon-colored VW Bug with mag wheels and rolled and pleated upholstery…a battered '62 Dodge Dart with bird shit on the hood and UCSB parking stickers piled up in one corner of windshield…" (109). In the book's appendix, one literary worker recounts heading to Matzhoh Momma's deli on Capitol Hill, where a song called *Disappear* was performed; patrons were asked to fill in lyrics, on a Mad-Libs-style worksheet that fueled the slightly changing versions of the song that appear throughout the novel.

But despite the lighthearted spectacle of this endeavor, Invisible Seattle also serves as an important critical voice in the city's growth. Founded in 1979, disbanded in 1993, the group (of which Wohlstetter was a key member) named itself after Italo Calvino's novel *Invisible Cities* (1978), and proceeded to create a mythology about Seattle. According to them, Visible Seattle was place where people lived, worked, and went on about their daily lives. Invisible Seattle, however, was a place where one could consort with "every great thinker and artist of all time" (Talking Stick TV, 2018). A slogan: "Every time you read a book, you enter Invisible Seattle" (Ibid.). Their activities ranged from putting "enemies of civic life" on trial to producing Surrealist or Situationist International-style maps of the city, to creating a literary bulletin board, IN.S.OMNIA, and installing public computer terminals at community hubs like the Elliott Bay Book Company and Wallingford Center, for people without computers to participate in writing experiments—a project that lasted about a decade and established them as progenitors of electronic literature.

Invisible Seattle was therefore clearly involved in a version of OA, but their interventions also extend the purview of such appropriations. The process of collective authorship undoubtedly drew various participants together over the course of a month, but it was not only a matter of spectacles in physical space—in the Seafirst plaza, on street corners, or in the Shaw Room at Seattle Center during Bumbershoot, where Scheherzade II's lights, bells, popcorn popper, and toaster making toast would be activated when literary workers drew large portions of the collective text into the story (Waldrop, 1984)—but also in mediatic and virtual space (these were, it must be remembered, different in the early 1980s), as well as in the collectively imagined space of the city. In the media, for example, a draft of the novel was read in full on KUOW, one of the local NPR affiliate radio stations, during a three-week stint later that year (Fig. 4). The IN.S.OMNIA project, in which users—"in.s.omniacs"—from across the country describe as a "digital coffeehouse, a place of refuge where they can be their various selves, a non-place on which to stand and survey other zones—books, billboards, brochures—where texts operate" (Wittig 1994, 2), strove to produce an alternative art to help them make sense of the world. As Rob Wittig, a literary worker from the novel writes: "Invisible Seattle's search for an alternate geography has produced the perfect non-place" (Witting, 1994: p. 7).[3] In "Voyage Through the Lobby," for example, the in.s.omniac PERSON begins a text that reflects on the entry portal to the bulletin board, the lobby, on March 21, 1984:

> I'm sure that everyone who has entered this lobby and has gone on into the board has seen, at one time or another, a slight, wiry figure in a baggy suit haunting the potted palms in the little alcove beyond the reception desk. He mechanically adjusts the crease in his trousers as he stares at the woodwork; or he carefully propels a smoke ring into the devastating path of the ceiling fan; or he folds a newspaper into crisp quartersize and raps it informatively with a few manicured nails, or he descends the length of his enormous watch-chain to pop open

[3]See Marc Augé (2009) for the genesis of the term "non-place" or Ian Buchanan (2005) for a Deleuzean interpretation of this idea.

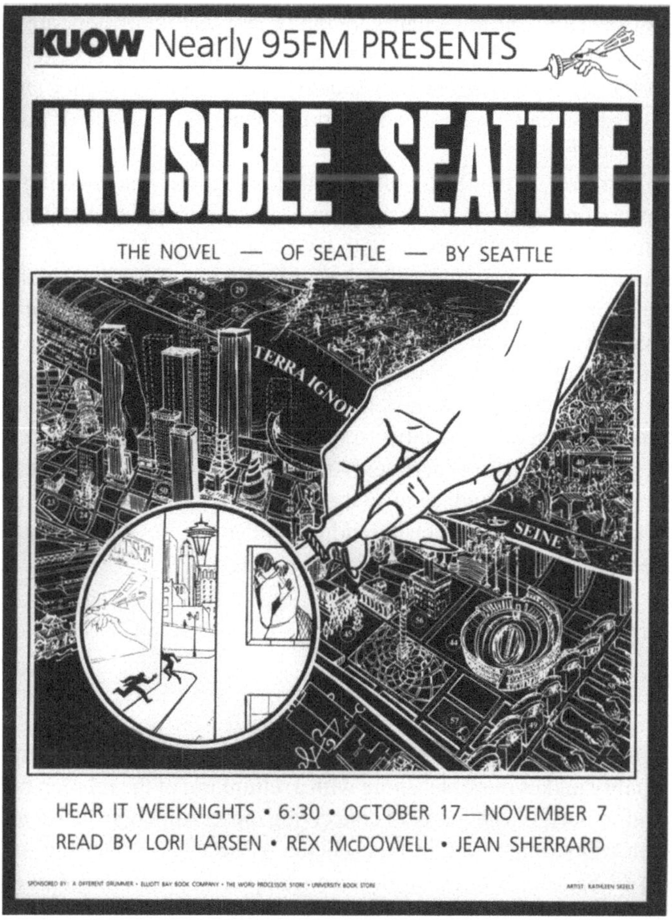

Fig. 4 Publicity poster for the radio broadcast of *Invisible Seattle. Art by Kathleen Skeels*

> his Swiss gold and cranes to see the revolving door, as though someone extraordinary was expected any minute (IN.S.OMNIA, 1985: p. 5).

Other in.s.omniacs—Lower Please, The Receptionist, and Joe—pick up the strand, creating new characters (a female companion, a receptionist) and sketching out the environment and characters' concerns over the next five months.

Finally, the Invisibles' rendition of Seattle subverts any mimetic understanding of Seattle. A mental map that could be attributed to one author or the overlapping mental maps of multiple authors, *à la* Kevin Lynch (1960), is shattered into a differential set of perceptions that are strewn across a plane—the book—that was constructed through the rules and questions posed by collective: a far northern suburban neighborhood, Meadowbrook, is posited alongside accounts of the historic downtown Camlin Hotel. The effect on the reader who is familiar with the actual layout of the city is one of destabilization: actual geography and view corridors are scrambled and a simultaneity that is rarely, if ever, experienced in real life is generated. More precisely, in the language of transindividuality, the individual reader and the population of Invisible Seattle—the Seattle lurking beneath the space of our everyday lies—emerge together.

To create the novel, to build it up like a cathedral, required establishing a new gesture, a new way of using various urban spaces—street corners, restaurants, plazas—over a sustained duration. Moreover, both the presentation of this work—in written form and over the airwaves—makes for the appropriation of mediatic and imaginary space, while the group's experiments with IN.S.OMNIA extended their reach into the nascent virtual realm. In the case of the novel, a new "I" and "we" can be thought and experienced, even when reading it 35 years after its construction. To read it today, for the first or the nth time, is to, by definition, emerge as part of a radical collective committed to re-envisioning the city. To reflect on the early promises of electronic literature, to see these playful experiments among anonymous users as creating a new form of sociality that exists adjacent to the pressures of production and capitalism, is to see once again modes of transindividuation that are not easily captured in the present era of capitalist realism.

Conclusion

In his final chapter, Read (2016: p. 249) writes, "It is possible to suggest that what transindividuality makes possible, along with a rethinking of the relation between the individual and society, is a new way of conceiving the relation between determination and liberation." Determination, in this schema, is best understood through Read's investigation of Stiegler's work on how consumption in contemporary capitalism structures the intertwined individuation process of both individuals and collectives. Liberation, then, is necessarily related, at least in part, to how we consume—or let's say use—space. However, my argument has focused not on the characteristics or qualities of the spaces themselves, but rather the necessity of appropriating any and all spaces for extended durations. The new gestures that emerge in cases of OA—that is, the willingness to occupy, to appropriate any-space-whatsoever, be it physical or virtual, in order to discuss, to learn, to dream, and to create—are a point of entry for constituting new political characters and alliances, or in Read's terms, new transindividuations. The extended duration of Red May as a teach-in, the construction of *Invisible Seattle*, or the decade-long experiment that was IN.S.OMNIA are examples of how we can consistently augment the tensions of capitalist reality, how we can push the transindividuation process in the direction that we want it to go: Toward a permanent vacation from capitalism.

References

Augé, M. (2009). *Non-Places: An Introduction to supermodernity*. London and New York: Verso.

Benjamin, W. (1969). *The work of art in the age of mechanical reproduction. Illuminations*. New York: Schocken Books.

Buchanan, I. (2005). *Space in the age of non-place, in deleuze and space*. Edinburgh: Edinburgh University Press.

Calvino, I. (1978). *Invisible cities*. New York: Harcourt Brace Jovanovich.

Citton, Y. (2012). *Renverser l'insoutenable*. Paris: Seuil.

Deleuze, G. (1994a). *Bergsonism*. Translated by Hugh Tomlinson and Barbara Habberjam. New York: Zone Books.

Deleuze, G. (1994b). *Difference and repetition*. Translated by Paul Patton. New York: Columbia University Press.

Deleuze, G., & Guattari, F. (1987). *A thousand plateaus: Capitalism and schizophrenia*. Translated by Brian Massumi. Minneapolis: University of Minnesota Press.

Fisher, M. (2009). *Capitalist realism*. Winchester, UK: Zero Book.

Haider, A. (2018). *Mistaken identity: Race and class in the age of Trump*. Paper presented at Red May 2018, Seattle, WA, May 3, 2018. https://www.redmayseattle.org/red-may-tv.

Hess, A. (2018). *The existential void of the pop-up 'Experience'*. The New York Times.

IN.S.OMNIA. (1985). *IN.S.OMNIA: Invisible Seattle's OMNIA*. Seattle: Function Industries Press.

Invisible Seattle. (1987). *Invisible Seattle: The novel of Seattle, by Seattle*. Seattle: Function Industries Press.

Lara-Hernandez, J. A., & Melis, A. (2018). Understanding the temporary appropriation in relationship to social sustainability. *Sustainable Cities and Society, 39*, 366–374.

Lynch, K. (1960). *The image of the city*. Cambridge, MA: MIT Press.

May, T. (1996). Gilles Deleuze and the politics of time. *Man and world, 29*, 293–304.

Red May. (2017). Red May 2017 Schedule (archived). https://redfoxseattle.wixsite.com/mysite/week3.

Read, J. (2016). *The politics of transindividuality*. Chicago: Haymarket Books.

Singh, N. P. (2017). *Race and America's long war*. Oakland: University of California Press.

Swenson, T. (2017). Red Videojoy. https://vimeo.com/240330508.

Talking Stick TV. (2018). Interview with Philip Wohlstetter. https://www.youtube.com/watch?v=Y8Etcgmd1RI.

Waldrop, H. (1984). Invisible Seattle: A novel idea. *PC Magazine, 6*(3), 198–202.

Wark, M. Molecular red: Theory for the anthropocene. London and New York: Verso.

Wittig, R. (1994). *Invisible Rendezvous*. Hanover, NH: Wesleyan University Press.

Wohlstetter, P. (2017). From Invisible Seattle to Red May: On the Passage of a Few People Through Rather Neoliberal Times. Paper presented at the Joanne Brown Symposium on Violence and its Alternatives, Vancouver, BC, April 2017.

Temporary Appropriation and Public Space: Assessing the CPTED Principle of Activity Support

Alexander Gomez-Torres

Abstract The potential for reducing crime throughout modifications to the built environment has been extensively argued and implemented over several decades within Crime Prevention Through Environmental Design (CPTED) knowledge. In this context, empirical evidence has shown that the presence of CPTED principles—such as Activity Support, Natural Surveillance and Territoriality—is related to a significant shift in the recurrence of crime. Nonetheless, the extent of how these principles work is still under research. This chapter assesses the role of activities related to urban design as contributor of the temporary appropriation of the public space for crime prevention purposes. The study presents a critical assessment of a case study: the Third Millennium Park in Bogota DC, Capital District of Colombia. The CPTED principle of Activity Support was assessed across the place using a ratio mapping methods to understand the relationships between public space and crime. The results unveiled a significant low level of activities and a tendency for using signage mainly as sets of instructions for mandatory purposes rather than a strategy for helping the temporal appropriation of the public space. The research chapter concludes that the urban design intervention played a positive role in the reduction of crime within the areas in the case study. Though, its concentration variously increased throughout the locations in the surrounding areas. Hence, implications of territoriality, crime displacement and the access to comprehensive crime data should also be considered in further research.

Keywords CPTED · Urban design · Public space · Activity support · Temporary appropriation

---

A. Gomez-Torres (✉)
University of Auckland, Auckland, New Zealand
e-mail: jgom498@aucklanduni.ac.nz

A. Melis et al. (eds.), *Temporary Appropriation in Cities*,
https://doi.org/10.1007/978-3-030-32120-8_6

Introduction

Over more than three decades in Latin America, and particularly in Colombia which is ranked amongst the world's most violent countries (Butchart & Mikton, 2014; The World Bank, 2017), urban renewal policy and urban design interventions have been used for recovering urban deteriorated neighbourhoods and reducing crime. In 1998 the country's intentional homicide per 100,000 measured United States with a rate of 5.56 or Australia and New Zealand with less than 2.0, while Colombia reported a critical ratio of 60.1 (The World Bank, 2017). The same year, the World Health Organization appointed "The Santa Ines-El Cartucho" a neighbourhood located near the National Capitol, as one of the most dangerous places in Latin America (Morris, 2011).

To solve this conflict, the city council of Bogota demolished the entire Santa Ines neighbourhood and developed a metropolitan public space, the Third Millennium Park. After the opening of this major urban design intervention, the presence of CPTED principles such as "Activity Support", "Access Control" and "Natural Surveillance" are noticeable and the recurrence of crime has variously shifted within the park. Such is the case of Intentional homicide, which was positively reduced from 175 cases (1999–2002) to less than 20 cases (2003–2007) (CEACSC, 2015). In regards to robbery, the reduction has been estimated over 70% after the urban design intervention, yet, it variously increased in the surrounding areas (Llorente & Rivas, 2005). This phenomenon has been arguably labelled as crime displacement. Despite this empirical evidence, the extent to which and how these activities within CPTED principles may be relevant in this non-Western urban context is far from conclusive.

Even though CPTED is accepted as an instrumental theory, methods for measurement from quantitative and qualitative approaches are still under research. As a result, despite CPTED being increasingly deployed as an urban design guideline, the way CPTED and its principles precisely deliver the expected outcomes is still far from completely understood. Although the urban design intervention may have improved aesthetical aspects of the renewed areas, the site is continuously reported as a place with low levels of human activity and uneven temporary appropriation. Within these considerations, this chapter aims to answer the research question: How relevant is the CPTED principle of Activity Support to understand the implications of temporary appropriation over the public space?

Environmental Crime Prevention, Urban Design and Public Space

In terms of environment–behaviour studies and design research, Environmental Crime Prevention may be considered a set of explanatory theories of the middle range (Moore, 1997) that aims to understand the presence of crime and the conditions and settings of the environment (Andresen, 2014: p. 30). After the 1970s, and primarily from a Western perspective, studies such as Crime Prevention and Control Through Environmental Engineering (CPTEE) (Jeffery, 1971), Defensible Space Theory (Newman, 1972), and the Broken Windows thesis (Kelling & Wilson, 1982), explored the relationship between built environment and crime. These studies are part of what is currently known as Environmental Crime Prevention wherein the idea of Crime Prevention Through Environmental Design (CPTED) emerged.

The term "CPTED" was first used in the studies by Dr Ray Jeffery in 1971; however, its principles and strategies are more aligned with the Defensible Space approach developed by Oscar Newman (Lersch, 2007). Despite the early studies in the 1970s, current research acknowledges various discrepancies in how CPTED is defined. Generally, the studies published in 1991 by criminologist Timothy D. Crowe are currently considered an important theoretical contribution for the consolidation of CTPED as an instrumental theory for crime prevention and urban design. Crowe's version states: "The proper design and effective use of the built environment can lead to a reduction in the fear and incidence of crime, and an improvement in the quality of life" (Crowe, 1991: p. 40).

CPTED Principles

Overall, CPTED researchers acknowledge at least six different principles: (1) Image Management, (2) Surveillance, (3) Access Control, (4) Target Hardening, (5) Legitimate Activity Support, and (6) Territoriality (Cozens & Love, 2015). Nonetheless, how these principles are denominated has changed over time. Current practice uses a various range of synonyms to refer to the same principle. For instance, Activity Support, also known as Activity Support Programme (Moffat, 1983) or Legitimate Activity Support (Cozens & Love, 2015). "Territoriality" is also denominated as "Territorial Reinforcement," "sense of proprietorship" or simply as "ownership." The excess of labels generates inaccuracy, making the concepts useless and creating a barrier for enabling CPTED as transferable knowledge (Gibson & Johnson, 2016), particularly within the context of urban planning and urban design.

The CPTED Principle of Activity Support

Activity Support, also known as Activity Support Programme (Moffatt, 1983) or Legitimate Activity Support (Cozens & Love, 2015) is the use of design features to deliberately increase the presence of "legitimate users" in a given area. The assumption of high levels of human activity as a wanted scenario to reduce the presence of crime is rooted in the concept of "eyes on the street" as appointed by Jacobs (1961). As a crime prevention strategy, it is built on the assumption that, as more activities that are "legitimate" are performed in a specific location, the fewer chances there are for a crime to happen (Cozens, 2014). Activity support is also a key concept to enhance territoriality, access control and surveillance (Cozens, 2014: p. 397). Furthermore, since the final aim of activity support is to "enhance surveillance" it might hardly be considered an independent CPTED principle but rather as a complementary one (Armitage, 2014, p. 723).

The application of this principle follows two key set of assumptions: the qualities that make a place considered to be safe, and the tendency of some activities to be unsafe by default (Crowe, 2013). In this sense, a safe place involves the presence of human activity with the highest opportunities for surveillance. Yet some daily commercial activities, like financial transactions, pubs or bars, may be categorised as unsafe (Cozens & Love, 2015). The allocation of such activities in suitable places is the very aim of activity support as a CPTED principle.

CPTED, Territoriality and Temporary Appropriation

From an Environmental Psychologist approach, Temporary Appropriation (TA) reflects the relationship between and individual (or group of people) and its surroundings (Pol Urrútia, 2002). In the context of public space, TA is also understood as the practice of multiple activities (Economy, Leisure and Sacralisation) beyond the original design programme (Lara-Hernandez & Melis, 2018). For instance, selling ice-cream along a sidewalk; skateboarding over a public stairs pathway; or promoting services or products over a public plaza. These type of spontaneous activities sometimes labelled as informal are considered a cultural reflection of a place and are assumed to improve the sense of belonging and identity within a given public space (Hernández Bonilla & Gómez Gómez, 2015; Portal, 2009).

The assumptions behind the concept of Temporary Appropriation are arguably aligned with the principle of Territoriality in the context of Environmental Crime Prevention. Territoriality is also known as the sense of proprietorship or simply ownership. Newman defined this principle, within Defensible Space Theory, as the capacity of the physical environment to create perceived zones of Territorial influence (Newman, 1972: p. 51). Territoriality has been argued to be the main aim of all CPTED principles (Crowe, 2013; Cozens, 2014). As a result, Natural Surveillance (Increasing eyes on the street) (Jacobs, 1961) and "Activity Support" (Increasing the levels of human activity) (Cozens & Love, 2015) contribute to reinforcing territorial responses.

Despite the similarities and differences within these notions, Temporary Appropriation unveils an important attribute of the built environment and the public space. As TA strongly vary from place to place and within different cultures, it raises implications behind building a comprehensive framework for a quantitative and qualitative assessment. Currently, the empirical evidence of TA is far from conclusive at least from the urban design and urban planning point of view.

Methodology

In order to answer the research question: "How relevant is the CPTED principle of Activity Support to understand the implications of temporary appropriation over the public space?" this methodology uses a mixed-research approach that is framed as an indirect non-participant observation through a case study: the Third Millennium Park in Bogota, Colombia. The methods deployed for this purpose have been conceived in three parts. The first part uses an archival research method, which is required for having the documents urban design from the original master plan. It also gathers the empirical evidence of crime related to the area of study. The second part uses mapping and a ratio map per thousand square metres method for the critical analysis of the urban design in terms of the CPTED principle of Activity Support. This enables the study to uncover the presence of activities within the physical designed features. The final part analyses the results from the urban design CPTED assessment and the implications for Temporary Appropriation alongside the shift of crime and delivers the findings of the research via logical argumentation (cultural/discursive).

The Case Study

With almost five centuries of urban history, the Colombian capital, Bogota, is one of the ancient cities in South America. For more than three hundred years the central city and the Santa Ines neighbourhood (located near the National Capitol) remained as a small colonial town; but, after the end of the war of a thousand days (1899–1903) at the beginning of the Twentieth century a rapidly and steadily migration wave from the countryside merged into the capital city producing rapid urban growth that occupied more than four thousand hectares in less than fifty years (Hudson, 2010; Pérez, 2000). Nonetheless, as the Santa Ines became an enclosed area surrounded by arterial roads; the traditional neighbourhood endured a steadily built environment deterioration with a progressive presence of criminal events such as homicide and drug micro-trafficking.

In 1998 the country's intentional homicide per 100,000 reported Colombia with a critical ratio of 60.1 (World Bank, 2017). The same year, the World Health Organization appointed "The Santa Ines-El Cartucho", as one of the most dangerous places in Latin America (Morris, 2011). After the WHO report, the City Hall demolished the entire neighbourhood in order to create a major public space; the Third Millennium Park (See Fig. 1).

In terms of Brantingham and Brantingham (1993), "The Third Millennium Park" may be considered as an urban design intervention in a Meso-analysis scale. This level of analysis might include intermediate units such a neighbourhoods, or subareas in a city. Table 1, describes the case study boundaries in a clockwise sequence.

Fig. 1 The third millennium park, case study. map from IDECA, bogota city

Table 1 Case study coordinates

Cardinal point	Urban limit
East:	La decima avenue
South:	The 6th street ave (former san agustin river)
West:	La caracas avenue (former 13 and 14 streets)
North:	The 9 th street

Mapping the CPTED Principle of Activity Support

In general terms, Activity Support (AS) aims at increasing the presence of people as a contributor to Natural Surveillance. In other words for supporting the assumption of "Eyes on the street" (Jacobs, 1961). Furthermore, some of the activities CPTED are also intended for enhancing Territorial Reinforcement (See Table 2), which may be

Table 2 Activity Support Methods

Method	Aim	Assumption
Comprehensive signage must be provided	Definition and delimitation of the public space	Reinforce Territoriality
Supporting Activities may be located near bus stops	Redirect human presence	Increases eyes on the street
Guarantee formal street vendors	Support activities over the public space	Increases natural surveillance
Mixed uses through day and night are advised	Improvement and concentration of activities by the public space 24/7	Increases natural surveillance

translated into developing a "Sense of ownership" (Cozens, Saville & Hillier, 2005). With these arguments, this methodology acknowledges the quantitative and qualitative aspects of the presence of activities in the case study. For the critical assessment of "Activity Support" as a CPTED principle, these research paper appraises the following interrelated assumptions.

The maps of "Activity Support" assesses within the CPTED arguments the extent of to which the signage may relate to a specific area. As a result, this map has been assembled as a "Ratio map" and aims to determine the location of all type of signs within zones and subzones of the study area. This is done through the three following steps:

- Counting (using google street view.
- Building a Ratio map per each thousand square metres.
- Qualifying signage and activities (information, identification, restriction).

Counting: Considering that the extension of the study area is more than 160,000 square metres, all signs has been located within a map of zones and subzones. The identification of these areas referred to the concept of "delimitation" and are part the notions behind the CPTED principle of Access Control. This is explored in a separated paper. Hence, the terms zone and subzones are used in this assessment for referring to the areas and sub-areas present within the limits in the case study (See Table 3).

The counting was made digitally through images and maps. Then the images were compared with DWG CAD digital files provided by the ERU (Bogotá Renewal and Urban Development Enterprise) and high resolution aerial photography of Bogota Capital District from IDECA Bogota Map. Finally, data of the urban design features are complemented with 3D captures from Google Street View from September 2012 (the most recent data available in Google for the location) (See Fig. 2).

Ratio Map: Depending on the number of signs present, each subzone has been quantified either as high, intermediate or low (see Table 4). The overall ratio per each thousand square metres was recorded on each subzone and graphically represent through a chromatic degradation. High Ratio: Areas with highest presence of activities (Light Yellow colour). Intermediate Ratio: Areas with medium levels of

Table 3 Zones and subzones in case study

Zone	Subzone	M_2	%
A	A1	9,377.94	5.80
	A2	17,729.59	10.97
	A3	8,368.36	5.18
B	B1	12,542.32	7.76
	B2	21,525.11	13.32
C	C1	13,583.78	8.41
	C2	18,545.76	11.48
Principal pedestrian areas		59,901.31	37.07
Total		161,574.17	100

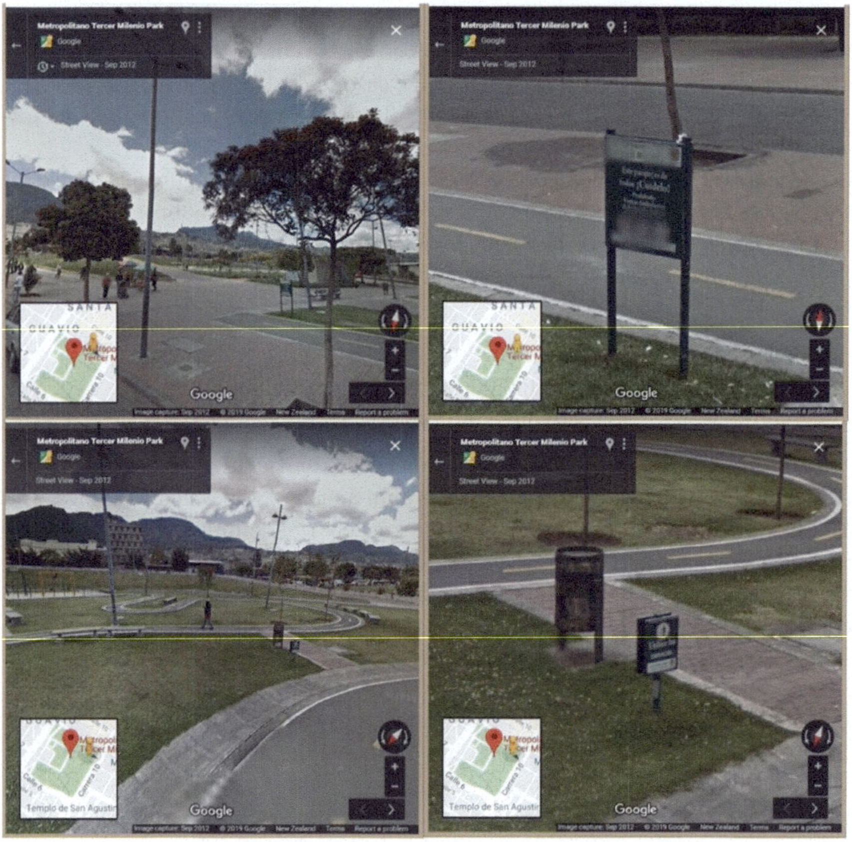

Fig. 2 Signage in the third millennium park. captured from google street view

Table 4 Urban design inventory in the third millennium park

Ratio	Signs/1000 M_2
Low	0.00–0.66
Intermediate	0.67–1.33
High	1.34–2.00

signage (Ochre Yellow colour). Low Ratio: These are areas with a recurrent deficit of signage scoring as the lowest (Dark Ochre colour).

Qualifying signage and activities: How much is too much with regard to signage and activities? In general terms, signage, as an urban design feature, provides a better understanding of the public space while it is treated and arranged as a whole rather than as a separate elements. Hence, in order to build a more comprehensive picture of its presence in the case study, signage and activities have been assessed with the following four complementary approaches. First, signage has been measured for unveiling as whether or not signs are located beside accessible sidewalks (See Table 5). Second, signs have been quantified in conjunction with facilities and activity generators (See Table 6). Third, signage has been assessed to determine how many signs are placed for the identification of places within the areas in the case study (Maps for location, information/identification, communicating single rules). Lastly, bus stops, formal street vendors and night mixed uses are discussed.

Findings and Discussion

Quantifying Activity Support Through Signage

Signage is currently a well-accepted and significantly explored practice within urban design and other built environment related disciplines. Generally, it is usually intended to give directions within an area, supplying information about activities or places. It may also be used for pedagogical purposes as a strategy for providing the historical, cultural, geographical background of a defined area. This is assumed to be a way of enriching a sense of place (PPS, 2008). Furthermore, in terms of Crime Prevention through Environmental Design, signage is also a means of providing a sense of proprietorship and therefore of supporting Territorial Reinforcement (Cozens, Saville & Hillier, 2005; Cozens & Love, 2015).

A quantification of signage, measured eighty five (85 UN) signs located in various places throughout the study area (See Fig. 3). Yet, a simple count may not be sufficient to determine just what 'comprehensive' is. Since the aims of Activity Support include the increased presence of people to support Natural Surveillance and Territoriality, it may be the case that it is not simply about discussing the quantity but the quality of the signage in relation to place. In this regard, characteristics such as location, frequency and proximity of signage are considered relevant within the terms of

Table 5 Quantity of signs located along accessible pedestrian networks

		A1	A2	A3	B1	B2	C1	C2	Total
Sub-zones	UN	13	11	7	6	13	3	4	57
	%	15.29	12.94	8.24	7.06	15.29	3.53	4.71	
Zones		36.47			22.35		8.24		67.06

Table 6 Number of signs placed in conjunction with facilities and activities generators

		A1	A2	A3	B1	B2	C1	C2	Total
Sub-zones	UN	7	3	5	2	4	1	3	25
	%	8.24	3.53	5.88	2.35	4.71	1.18	3.53	
Zones		17.65			7.06		4.71		29.41

Fig. 3 Signage inventory in the map of the third millennium park

CPTED. For these reasons, apart from place-based quantitative measures, a more qualitative approach assessment after a ratio map [per sub-zone] has been added to this appraisal.

Ratio Map: The overall ratio per each thousand square metres shows that the highest presence of signage is within zone A. This is despite that the sub-zone A2 ranges along the lowest, which may relate to the fact that it is mainly an open green area. Sub-zones B1 and B2 perform within the same intermediate range (from 0.67 to $1.33/1000^{M2}$). Lastly, the entire zone C, has the lowest ratio of signage presence (See Fig. 4). Although, these results, raise the question: How much is too much with regard to signage and activities?

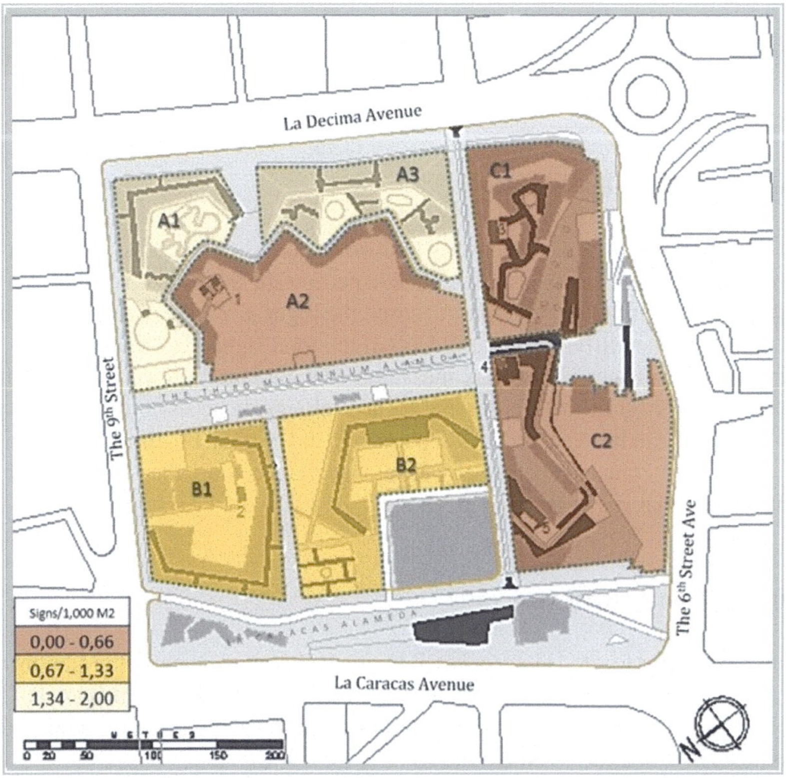

Signage Quantification	A1	A2	A3	B1	B2	C1	C2
Subtotal per Sub-zones	18	11	15	10	17	7	7
Total per Zones	44			27		14	
TMP Total Signage	85						
Quotient	9,38	17,73	8,37	12,54	15,08	13,58	19,18
Signage Ratio per / 1.000^{M2}	1,92	0,62	1,79	0,80	1,13	0,52	0,36

Fig. 4 Signage ratio per 1000 m^2 in the third millennium park

Qualifying Signage and Activities

The following section explores the answer to the question: How much is too much with regard to signage and activities?

Signage and Accessible Sidewalks

Overall, the case study measured fairly high with more than sixty seven per cent (57 UN) of signs being located beside accessible pedestrian sidewalks. With an average of thirty six per cent (31 UN), zone A scored the highest. Zones A and B combined represent more than fifty per cent (50 UN) of the total signage (See Table 5).

However, sub-zones C1 and C2 measured as significantly lower with less than ten per cent (7 UN) of the signs being located alongside accessible sidewalks. As discussed earlier in this chapter, in comparison to the original blueprint of the Master Plan, the C2 sub-zone is only partially developed; hence, more than forty per cent of the place has been left as an open green area.

Signage Within Activity Generators

In contrast to signs located along accessible sidewalks, signage placed in conjunction with facilities and activity generators measure considerably lower with a total of less than thirty per cent (25 UN). Activity Generators is a general term to highlight the importance of amenities in an urban area.

Zone A remains as that with the highest score with slightly less than eighteen per cent (15 UN) of the case study. In contrast, the remaining zones measured even less with less than ten per cent in zone B (6 UN) and less than five per cent (4 UN) in zone C (See Table 6). This outcome supports the previous findings concerning the recurrently low level of activity support measures throughout the park.

Signage Identifying Places Within Places

The identification of places within places (zones and sub-zones) is done through the three following tasks. First, measuring the number of signs that function as maps to provide location. Second, counting the amount of signs that provide information and identification of a specific area within the park. Lastly, by quantifying which of the signs have been placed to communicate single rules (for instance, "dog status"; "smoke free area"; "alcohol restriction") (See Table 7 and Fig. 5).

The results of signage identifying places within places scored as being remarkably uneven. Despite the promotional map that is available online on the City of Bogota's website (http://idrd.gov.co/sitio/idrd/Documentos/plano%20tercer%20milenio.jpg),

Table 7 Signage identifying places within places

Sign task		A1	A2	A3	B1	B2	C1	C2	Total
Maps for location	UN	0	0	0	0	0	1	1	2.00
	%	0.00	0.00	0.00	0.00	0.00	1.18	1.18	
		0.00			0.00		2.35		2.35
Information identification	UN	7	2	6	1	4	5	4	29.00
	%	8.24	2.35	7.06	1.18	4.71	5.88	4.71	
		17.65			5.88		10.59		34.12
Communicating single rules	UN	11	9	9	9	13	1	2	54.00
	%	12.94	10.59	10.59	10.59	15.29	1.18	2.35	
		34.12			25.88		3.53		63.53
							UN		85
							%		100

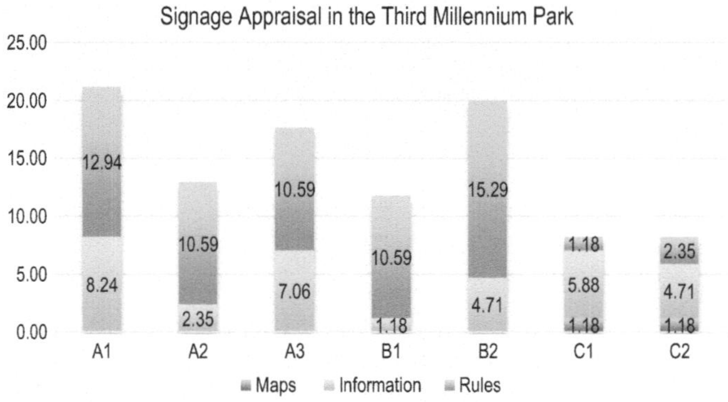

Fig. 5 Signage qualification in the third millennium park

locating nine (9 UN) strategic main access points for accessing the park, the assessment report has only two (2 UN) entries placed within the sub-zones C1 and C2 (See Fig. 5).

Signage for information and identification remains average throughout the area with thirty four per cent (29 UN) of the total signage; although, more than sixty per cent (54 UN) of the signs in the Third Millennium Park are placed for communicating single rules (See Fig. 5). This outcome reflects a tendency for using signage mainly as sets of instructions for mandatory purposes rather than supporting temporary appropriation and therefore improving the experience of being in a public space.

Street Vendors, Bus stops, and Mixed uses

The final attribute map for Activity support assessment in the Third Millennium Park has purposely been left to illustrate three more CPTED measures. First: "Street vendors may be formally and strategically placed to improve (eyes on the street)"; second: "Supporting activities may be located near bus stops or public transport spots"; lastly: "Mixed usage throughout day and night within the public space is highly recommended (coffee shops, cinemas, galleries)" (See Fig. 6).

The general measurement of "Activity Support" as a CPTED principle in the case study implies a certain degree of challenge. Though, the measurement of street vendors, bus stops and mixed use throughout the study area might be considered as being extremely difficult to achieve. The captures from the Google Street View include four locations with the presence of formal street vendors, strategically located by Bogota city council. Two of them are situated along "La Decima Avenue"; one at the junction between "La Caracas Avenue and the 6th Street Avenue; the last one is in front (North façade) of the "National Institute of Legal Medicine" (See Figs. 6 and 7).

Arguably, the number of formal street vendors may be low. Yet, when compared to the number of informal street vendors, the quantity may be considered irrelevant.

Fig. 6 Activity support: street vendors and bus stops map in the third millennium park

Usually located along "La Decima Avenue", the social phenomenon of informal street vendors has been a recurrent concern for several years for both local government and the police. As argued, Temporary Appropriation reflects the conflict between the activities included in the original design programme against unplanned activities such as those for informal economic gain of the public space.

These types of unauthorised activities are not per se related to crime. Nonetheless, in the case of the Third Millennium Park, informal sale activities are recurrently linked to illegal trading and smuggling, commonly known as the black-market, and

Fig. 7 Formal street vendors (Above #1 L & #2 R) and Informal Street vendors (below #3 L & #4 R). captured from google street view

to various types of robbery including pickpocketing and simple assault. From these considerations, this research highlights the importance of the context in measuring CPTED in the built environment. Some of the issues that present in the public space go beyond the scope of urban design and require interdisciplinary discussion.

Concerning supporting activities located near the bus stops or public transport areas, the case of the Third Millennium Park reflects an isolated spectrum. The only two bus stops in the area of the entire case study remain as secluded signs without the minimum requirement of an appropriately designed bay in which buses can stop. Despite the fact that these bus stops are located along arterial roads with a significant volume of traffic travelling at a considerable speed. Additionally, the nearest transport hub "Transmilenio transportation system" is located outside the public space area in the middle of "La Caracas Avenue". Access is restricted solely to the traffic lights located on the corner between "La Caracas Avenue and the 6th Street Avenue. The second possible access is placed two hundred metres away from these traffic lights in a northerly direction along "La Caracas Ave." (See Figs. 6 and 8).

Finally, in regard to "Mixed use throughout day and night within the public space is highly recommended (coffee shops, cinemas, galleries)"; the overall picture of the "Activity Support" remains quite blurred. The urban design inventory does not show any evidence of cinemas, galleries or even coffee shops throughout the study area. Of even more concern is the fact that the Third Millennium Park closes at 6:00 pm including at weekends, and any possible night-time activity is legally restricted. Therefore, Temporary Appropriation during these hours is hardly perceived.

Fig. 8 Bus stops in the third millennium park. captured (above #5 L & #6 R; below # 7 left) from google street view and IDECA (below right)

Conclusion

This chapter has assessed the relevance of the CPTED principle of "Activity Support" to understand the implications of temporary appropriation over the public space. For this purpose, "The Third Millennium Park", located within the city centre of Bogota in Colombia, was chosen as a case study. For answering the research question, the investigation was undertaken using a "Ratio Map Method" and captures from Googles Street View assessing the presence of the principle of Activity Support over the urban design features. Implications about Temporary Appropriation were also uncovered as a result.

Based on the assessment of signage, bus stops, formal street vendors and night activities, there is significant and therefore, a concerningly reduced quantity of activities to support temporary appropriation and the presence of people in the study area. First, there is also an evident incidence of signage, used mainly for displaying regulations and restrictions, rather than contributing to the enrichment of experiences within the public space. Bus stops are no integrated with the place activities and main transport hubs are located in the middle of arterial roads. Furthermore, the presence of formal street vendors may be considered irrelevant in contrast with the presence of informal economic activities over the public space. What is more, night activities are restricted after the sunset since the park closes after six pm.

To sum up, this outcome represents an urban design area with a significant deficit of Activity Support from a CPTED perspective and uneven temporary appropriation linked mainly to informal economic practices. Regarding the shift of crime after the urban design intervention, the reduction of violent crime against persons, "Intentional Homicide and Robbery", is considered positive. Nevertheless, the presence of these two types of crime has increased throughout various locations and concentrations in the surrounding areas. This phenomenon has been commonly argued from two opposing viewpoints; crime displacement and crime placement. These considerations help to the discussion of considering crime within the boundaries of time, victims, laws and place and contribute to the enrichment of our understanding of the presence of crime following the physical intervention of the built environment.

References

Andresen, M. A. (2014). *Environmental criminology: Evolution, theory, and practice*. Routledge.

Armitage, R. (2014). Crime prevention through environmental design. In *Encyclopedia of criminology and criminal justice* (pp. 720–731). New York: Springer.

Brantingham, P. L., & Brantingham, P. J. (1993). Nodes, paths and edges: Considerations on the complexity of crime and the physical environment. *Journal of Environmental Psychology, 13*(1), 3–28.

Butchart, A. & Mikton, C. (2014). Global status report on violence prevention.

CEACSC. (2015). Centro de Estudio y Análisis en Convivencia y Seguridad Ciudadana. [Centre for Study and Analysis in Communal Living and Citizen Safety]. Alcaldía Mayor. Bogota, D.C.

Cozens, P. (2014). Guidelines for design against crime think crime!: Using evidence, theory and crime prevention through environmental design (CPTED) for planning safer cities. G Quinns Rocks, Western Australia, Praxis education.
Cozens, P., & Love, T. (2015). A review and current status of crime prevention through environmental design (CPTED). *Journal of Planning Literature, 30*(4), 393–412.
Cozens, P. M., Saville, G., & Hillier, D. (2005). Crime prevention through environmental design (CPTED): A review and modern bibliography. *Property Management, 23*(5), 328–356.
Crowe, T. D. (1991) National crime prevention institute (University of Louisville). In *Crime prevention through environmental design: Applications of architectural design and space management concepts*. Boston, MA: Butterworth-Heinemann.
Crowe, T. D. (2013). *Crime prevention through environmental design*. In T. D. Crowe & L. J. Fennelly (eds.), 3rd edn. Amsterdam: The Netherlands, Elsevier.
Gibson, V., & Johnson, D. (2016). CPTED, but not as we know it: Investigating the conflict of frameworks and terminology in crime prevention through environmental design. *Security Journal, 29*(2), 256–275.
Hernández Bonilla, M., & Gómez Gómez, S. (2015). Conflicto espacial, exclusión y espacio público en la centralidad urbana de Xalapa, Veracruz. *Revista Brasileira de Gestão Urbana, 7*(3), 281–294. [Online] Available from: https://doi.org/10.1590/2175-3369.007.003.ao04.
Hudson, R. A. (2010). *Colombia: A country study*. Federal Research Division Library of Congress. [Online] Available from: https://www.loc.gov/item/2010009203.
Jacobs, J. (1961). *The Death and life of great American cities*. [Online] Available from: https://doi.org/10.2307/794509.
Jeffery, C. R. (1971). Crime prevention through environmental design. *Criminology, 10*(2), 191–191. [Online] Available from: https://doi.org/10.1111/j.1745-9125.1972.tb00553.x.
Kelling, G. L., & Wilson, J. Q. (1982). Broken windows: The police and neighborhood safety. The Atlantic.
Lara-Hernandez, J. A. & Melis, A. (2018). Understanding the temporary appropriation in relationship to social sustainability. *Sustainable Cities and Society, 39C*, 1–14. [Online] Available from: https://doi.org/10.1016/j.scs.2018.03.004.
Lersch, K. M. (2007). *Space, time, and crime*. Carolina Academic Press.
Llorente, M. V. & Rivas, Á. (2005). La caída del crimen en Bogotá: una década de políticas de seguridad ciudadana. *Seguridad ciudadana: Experiencia y desafíos*. 311–341. [Online] Available from: https://doi.org/10.1007/s15010-011-0129-1.
Moffatt, R. E. (1983). Crime prevention through environmental design-a management perspective. *Canadian Journal of Criminology, 25*, 19.
Moore, G. T. (1997). Toward environment-behavior theories of the middle range. *Toward the Integration of Theory, Methods, Research, and Utilization* (pp. 1–40). Boston, MA: Springer.
Morris, I. (2011). En un lugar llamado El Cartucho. *Bogotá: Instituto Distrital de Patrimonio Cultural*. [Online] Available from: http://idpc.gov.co/publicaciones/?product=en-un-lugar-llamado-el-cartucho.
Newman, O. (1972). *Defensible space: People and design in the violent city*. London, UK: Architectural Press.
Pérez, A. (2000). *Bogotá y Cundinamarca, expansión urbana y sostenibilidad*. Bogotá: CAR, Internacional Ltda.
Pol Urrútia, E. (2002). El modelo dual de la apropiación del espacio en Mira, R., Sabucedo, JM y Romay, J. *Psicología y Medio Ambiente. Aspectos psicosociales, educativos y metodológicos*, 123–132.
Portal, M. A. (2009). Las creencias en el asfalto. La sacralización como una forma de apropiación del espacio público en la ciudad de México. *Cuadernos de Antropología Social, 30*, 59–75.

PPS. (2008). *Signage*. 2008. Project for Public Spaces. [Online] Available from: https://www.pps.org/article/signage-guide. Accessed March 26, 2019.

The World Bank. (2017). *Intentional homicides (per 100,000 people)*. 2017. UN office on Drugs and Crime's International Homicide Statistics database. [Online] Available from: https://data.worldbank.org/indicator/VC.IHR.PSRC.P5. Accessed March 11, 2019.

Temporary Appropriation of Public Spaces: The Influence of Outdoor Comfort

Daniele Santucci, Ata Chokhachian and Thomas Auer

Abstract The term appropriation in general definition, is known as a temporary phenomenon that implies a dynamic process (human activities) of interaction between the individual and its surroundings (built environment). The term in generally used to illuminate the connection between people and public spaces in social science; however, the footprint of environmental studies is underestimated. This is not because our thermal sense of a place is inconsequential or superfluous, but simply because our scientific understanding of human thermal perception in the outdoor context is not yet at the stage where it can be purposively engineered and managed. This chapter intends to focus on the Temporary Appropriation (TA) happening in public spaces to give the citizens the right to fully use and manage their everyday life within the urban environment. This approach highlights the necessity of adapting human comfort to the needs of a city to compensate sudden and unexpected changes due to urban microclimate. In the context of the enormous urbanization phenomena occurring all over the world, the relevance of this issue increasingly gains importance. There is still an open question, if the daily public life or use of urban spaces can be acknowledged as appropriation or not. If yes, is this temporary or permanent? The concept of appropriation is closely tied to the sense of identity and comfort as well, besides temporal factors. This chapter will conclude supporting the idea that the temporary appropriation takes place in urban public spaces in which individuals or group of people feels culturally and environmentally identified.

Keywords Temporary appropriation · Public spaces · Human-centered approach · Outdoor comfort

Introduction

Appropriation has got diverse meanings and interpretations in different disciplines where in urban studies can be initiated as one of the informal approaches to characterize "Everyday Urbanism". Everyday Urbanism (Ahearne, 2010; Certeau, 1984;

D. Santucci (✉) · A. Chokhachian · T. Auer
Technische Universität München, Munich, Germany
e-mail: daniele.santucci@tum.de

A. Melis et al. (eds.), *Temporary Appropriation in Cities*,
https://doi.org/10.1007/978-3-030-32120-8_7

Speaks, Crawford, Mehrotra, & Kelbaugh, 2005) is referring to everyday life without the necessity of creating ideal environment. This means day-to-day activities does not seek to transform physical structures, and for that reason they take place in the already existing urban environment, along with other urban ideals (Liabäck Löwstett, 2018). In addition, everyday urbanism does not seek to transform places through comprehensive masterplans (Speaks et al., 2005) however, it emphasizes the micro sphere of the public realm (Haas & Olsson, 2014), an informal approach characterized by a bottom-up structure of grass-root initiatives. The debate is, even if the daily life (everyday urbanism) does not mean to create ideal environments, the quality of public spaces can have an impact on how people appropriate (not only temporary) modify, occupy or use public ground.

For Lefebvre (Lefebvre & Nicholson-Smith, 1991), appropriation is a function of lived space, that is, "dominated, hence passively experienced space, which imagination seeks to change and appropriate." Lived space is a place of passion where the user/inhabitant encounters resistance, which inevitably involves conscious action and struggle. In addition to community participation, appropriative strategies enable interdisciplinary collaborations that might inform, as Lefebvre noted, the "production of new spaces".

Most of the definitions on temporary appropriation considers environment as static phenomena which is not necessarily needed to be modified by the user however the dimensions and qualities of environment can have a direct effect of how people use or occupy public spaces. One of the main influencers on how people use public spaces is the outdoor comfort which is defined as local microclimate on pedestrian level influenced by environmental and physical structures of the city.

Climate is the primary factor that allows the use of public space. In particular microclimate, which describes the condition that human beings continuously experience through their senses. As such, it has a fundamental influence in giving quality to the urban experience. Walking is the primary way of appropriation.

In *The Practice of everyday life*, Certeau (1984) states:

> The ordinary practitioners of the city live "down below," below the thresholds at which visibility begins. They walk—an elementary form of this experience of the city; they are walkers, Wandersmänner, whose bodies follow the thicks and thins of an urban "text" they write without being able to read it. These practitioners make use of spaces that cannot be seen; their knowledge of them is as blind as that of lovers in each other's arms.

The act of walking to the urban system is like what the speech act to the language. At the most elementary level, it has a triple "enunciative" function: first, it is a process of appropriation of the topographical system on the part of the pedestrian (just as the speaker appropriates and takes on the language); second, it is a spatial acting-out of the place (just as the speech act is an acoustic acting-out of language); and third, it implies relations among differentiated positions, that is, among pragmatic "contracts" in the form of movements (just as verbal enunciation is an "allocution," "posits another opposite" the speaker and puts con-tracts between interlocutors into action). It thus seems possible to give a preliminary definition of walking as a space of enunciation.

> First, if it is true that a spatial order organizes an ensemble of possibilities (e.g., by a place in which one can move) and interdictions (e.g., by a wall that prevents one from going further), then the walker actualizes some of these possibilities. In that way, he makes them exist as well as emerge. But he also moves them about and he invents others, since the crossing, drifting away, or improvisation of walking privilege, trans-form or abandon spatial elements. He thus creates a discreteness, whether by making choices among the signifiers of the spatial "language" or by displacing them through the use he makes of them. He condemns certain places to inertia or disappearance and composes with others spatial "turns of phrase" that are "rare," "accidental" or illegitimate. But that already leads into a rhetoric of walking.

Following this premise and considering walking as a temporary appropriation of public spaces, cities are walkable—healthy, safe and vital - if they are comfortable. Therefore Microclimate most drastically affects cities, the places where society, economy and environment converge and where walkability, comfort and health are the most substantial indicators for appropriation, and consequently, of quality of the public realm. In fact, particularly cities and citizens experience the consequences of such phenomena, resulting in a high amount of challenges for the public space having serious impacts on the everyday lives and well-being of hundreds of millions of people around the globe. With different impacts from place to place.

The crucial task of research in outdoor thermal comfort, generated by micro-climatic conditions, consists of understanding and describing the correspondence between the artificial materiality of the "microclimate as human artifact, even though it appears to be a natural, nonmaterial and physical phenomenon" (Roesler & Kobi, 2018). Outdoor thermal comfort in urban spaces is fundamental to spatial appropriation as a result of the architecture and landscape architecture.

The research area of outdoor comfort, which makes use of computation, data mining and data analysis, considers the human itself as the inseparable dimension of design.

Design is one of the most powerful ways to favor some groups of people and disenfranchise others. It is therefore one of the most direct tools for affecting change and to create a more equitable world.

The intersection of climate and design, where the implications of research in the field have the potential to highlight inequities and furnish inclusive alternatives. Contemporary cities offer numerous examples of exclusionary design. Climate, in particular, has far-reaching effects on poor and under-served social groups. In this context, demographically diverse insights and broad-scope thinking can influence design decisions and increase equity at scales ranging from the super-local (the building) to the neighborhood (urban planning) to global policy.

Methodologically, there are two main fields to evaluate and predict comfort conditions in cities, simulation oriented and data integrated approach. Due to availability of powerful computational systems and cloud computing resources, nowadays running complex and multiscale models is an issue any more. This is where simulation-based methods to assess microclimate is getting more and more to the center of attention however our tools are not fully there yet to model every geometrical, physical and physiological behavior of microclimates. Furthermore, usually simulation-based approaches underestimate the human dimension. This is where the data integrated model can fill the gap. There are several limitation also in the data domain where the resolution and quality of the data is one of the main players. Besides the limitations, the sensing approach toward microclimate evaluation is based on the method that brings human to the focal point (human-centered approach) keeping in mind that nowadays availability of data is not the concern anymore, however it is more about methods and techniques to map diverse layers of data as a key for predicting urban phenomena. Furthermore, trying to provide a vision for the cities of the future, aiming to create a vibrant and healthy environment for people, which is able to mitigate thermal stress and also prepared to recover from more frequent and more extreme climatic phenomena (Chokhachian, Santucci, & Auer, 2017). The idea of approaching the problem from human-centered perspective studies the phenomena of microclimate in higher resolution in terms of human interactions with climatic conditions and urban structures. Furthermore, this approach works as bottom-up system placing human at the center of responsiveness considering needs, impacts and influences to start addressing solutions. The method is able to quantify self-regulating location-based design recommendations in different contexts.

Climatewalk, Sensing the Environment

As discussed before, one of the main challenges of data integrated methods is the availability and quality of datasets since most of them by default are not gathered specifically for human comfort evaluations. So one of the effective ways is to collect the data in diverse domains with knowing the hypothesis and the question that the data is going to address at the end. In this regard, the developed methodology (ClimateWalk) is specifically designed to measure environmental factors and human thermal behavior in urban spaces by relating individual and subjective responses to environmental conditions adaptation. The objective of the ongoing research, is to provide thermal understanding of the responses of pedestrians when walking in outdoor environment (temporary appropriation) using a georeferenced method for monitoring and mapping of microclimate (Chokhachian, Ka-Lun Lau, Perini, & Auer, 2018) and a longitudinal survey to obtain the thermal responses of pedestrians, for improving the "climatic knowledge" of the urban context. The study is relevant to the topic of urban density and outdoor comfort understanding the mutual impacts and bringing the human factor to the center of attention. Figure 1 shows the graphical setup of the methods merging environmental, physiological and psychological domains to have a human centered approach to monitor and map microclimate and outdoor comfort in cities. The novelty of our workflow is the combining existing techniques for understanding the complex interplay that determines confluence in a specific place and to put into evidence the effects of urban morphology on citizens. In particular, we can measure transient conditions from outdoor to indoor, and highlight the implications, on indoor conditioning measures and strategies for climate mitigation.

Neutral temperatures are often defined as a standard for outdoor thermal comfort and microclimate that contributes to a more sustainable outdoor environment

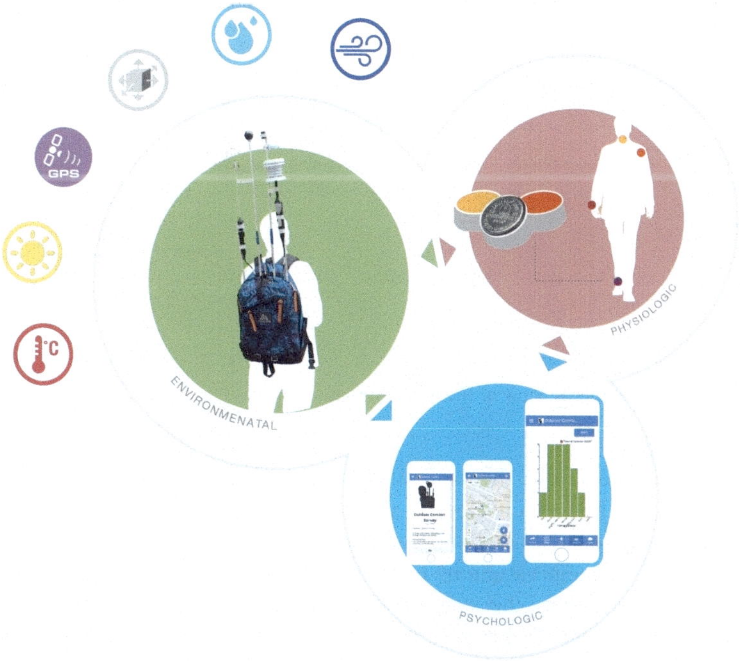

Fig. 1 The workflow of the ClimateWalk experiments, merging sensing domains of environmental, physiological and psychological data

(Johansson & Emmanuel, 2006). However, most of these studies emphasize on outdoor comfort as a "static" phenomenon, which is determined by buildings and other urban structures. Transient conditions are experienced when people move between spaces and are situated in spaces with a wide range of environmental conditions (Hensen, 1990) including step changes, temperature drifts, and cyclic variations influence thermal sensation and comfort. Pedestrians tend to adjust their behavior or walking routes in order to achieve better thermal comfort (Dear & Brager, 1998). The ClimateWalk experiments are designed to evaluate subjective behavior of people in transient conditions. The methodology is applied in several cities like Munich, Seville, Hong Kong, Rome, Singapore, Malaga and Barcelona. Here we present two examples from Rome and Singapore with very different climate zones and behavioral expectations to highlight the ranges.

Case of Rome

During the experiment that we carried out in Rome on June 11th and 12th 2018, between 13:00 and 15:00, we focused on the effects of urban morphology on thermal sensation and comfort (Fig. 2). Besides measuring UTCI with the sensing device,

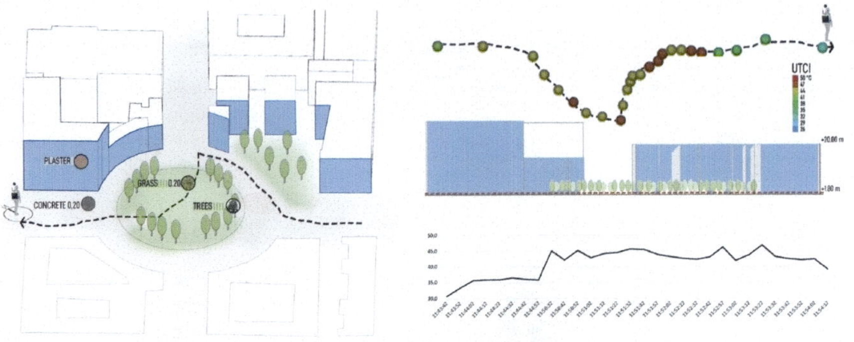

Fig. 2 Section and isometric view of the walking route in Prati

we analyzed the street ratio, the materials and green infrastructure in three different neighborhoods: the historic city, Prati—a turn of the century grid structure—and Torrevecchia, a social housing neighborhood in the northeastern periphery. In particular we assessed the frequent variations that occur when walking in the dense historic urban environment, where the canyon size and orientation varies often determining frequent physiological changes, and subsequently, a fluctuation in perception. Because of their material diversity and complex morphology, urban areas are characterized by varied microclimates across very small spatial and temporal scales. Furthermore, in public spaces, people are exposed to varying conditions due to their navigation across the spaces. Particularly in this transient conditions, thermophysiological regulation mechanisms set into maintain a stable body temperature, and people experience continuous substantial changes in terms of thermal sensation and thermal comfort. As it has been suggested earlier by Cabanac (1992) and Parkinson, de Dear, & Candido (2012), a perception of thermal pleasure does predominantly occur in transient states, because displeasure must occur in order to experience pleasure (Parkinson et al., 2012).

Case of Singapore

During this specific experiment in Singapore, the previous ClimateWalk setup has been combined with physiological measures, through precise and continuous skin temperature measurements. The experiment was designed to understand interdependencies of microclimate conditions within varied range of daily activities integrating the topic of adaptation in outdoor spaces.

The novelty of the present study is the amalgamation of three different measures in an outdoor environment:

- Microclimate georeferenced field measurements at the human scale assessed with the UTCI;

- Subjective thermal perception, recorded by an individual survey with designed App;
- Physiological response assessed by skin temperature measurements with iButtons.

The experiment was conducted in Singapore on July 25th and 26th 2018 and started at 11 a.m. and ended at 1 p.m. The days had different climatic conditions: the 25th was a cloudy day with reoccurring rain and showers (maximum air temperature during the walk: 30.9 °C), the 26th a sunny hot day without cloud coverage and precipitation (maximum air temperature during the walk: 34.9 °C). The participants were exposed to varying outdoor conditions in a dense urban context, simulating a common life experience, when moving from conditioned indoor spaces to outdoor public spaces and then again into conditioned commercial spaces. Subjects met in a Subway station where they were equipped with wireless skin temperature sensors, while they were exposed to a 30-minute 23 °C preconditioning period (65% RH) at the beginning of each experiment. Following this, the walk started. Every 10 min the subjects were asked to do a survey about thermal perception. The walk had a moderate speed and ended around 1 h after in an outdoor/indoor exchange environments. The experiments were conducted during consecutive mornings. This was done in consideration of circadian rhythms in core temperature, as well as the influence of prevailing weather on expectations and adaptation. The findings show that on the hotter day, conditions are drastically different: the equivalent temperature difference between indoor and outdoor is high (up to 7 K) contributing to a higher fluctuation of the mean skin temperature. In tropical climate, clear sky conditions contribute to a higher influence of the built environment, since the variations of direct radiation highly affect the UTCI values. While with overcast conditions, the contribution of the variation of the urban environment is less relevant on the UTCI values. In general, the most evident phenomenon we could recognize is that on the cooler day (July 25th) the mean skin temperature of women is higher, and on the warmer day (26th) it is lower than the male subjects' average (Fig. 3).

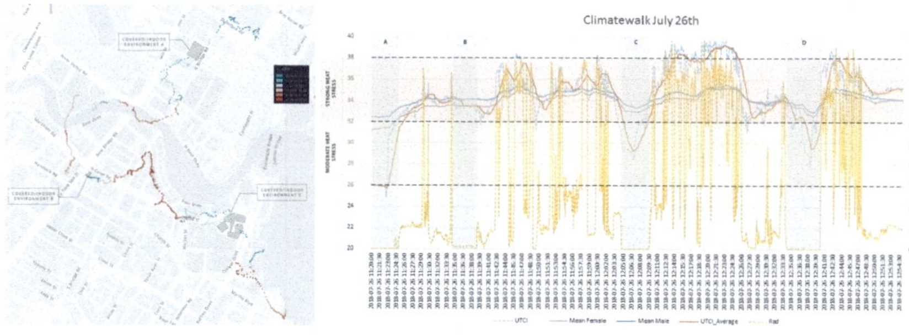

Fig. 3 The microclimate map of the route (left), UTCI measure and skin temperature of the subjects (right)

Thermal Perception and Appropriation

Thermal comfort, regardless of environmental parameters and energy balance, is still largely a state of mind. However, this perception is definitely influenced by environmental, heat and mass transfer variables. In most of the thermal comfort studies, the general approach is to confront numerical thermal comfort estimations with psychological metrics. In these methods, the level of comfort is often characterized by thermal sensation and thermal pleasure scales.

Embeded in the ClimateWalks experiment, we use a 7-point ASHRAE thermal sensation scales to assess subjective responses of the objects. The responds are collected with designed app on the predefined locations embedded with GPS trackers to trigger exact locations of the survey through the route. In case of Rome during our walks from Piazza del Popolo in Rome through the historic urban texture (Fig. 4), we recorded the variations in thermal comfort using the UTCI metric and the thermal response of the participants. The data were plotted into the maps and in the street section, to highlight the influence of the urban morphology (Sky view factor SVF, ratio, intersections, trees, canyon size, orientation, length, etc.). Previous studies

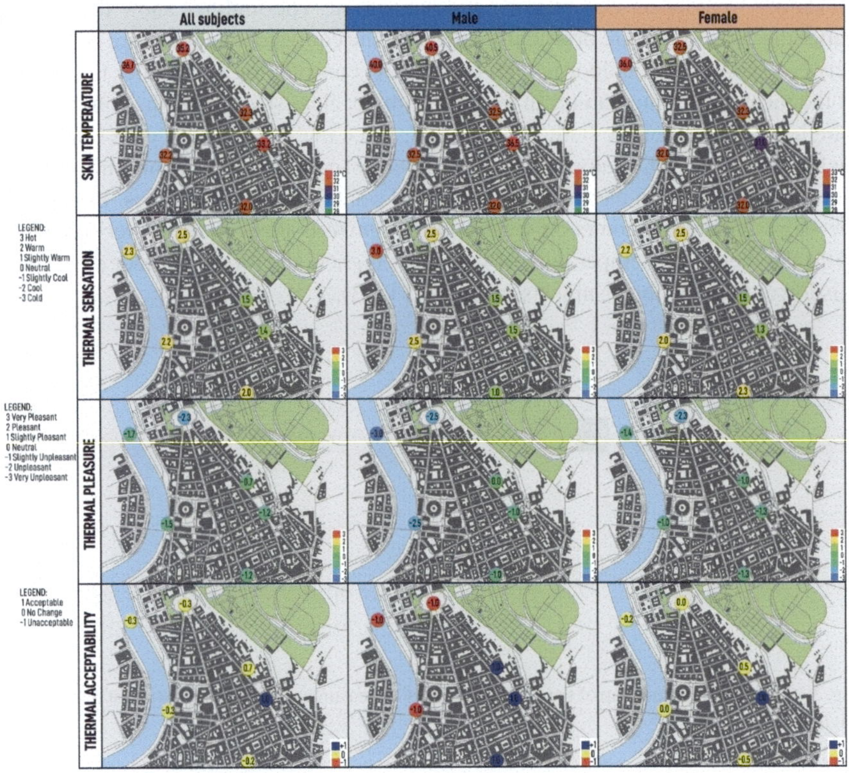

Fig. 4 Thermal sensation and Thermal pleasure maps of ClimateWalks in Rome

have illustrated to which extent pedestrians respond to the variability of the urban environment. The correlation between the SVF and the frequency of pedestrian activity along a street segment shows strong relations between the variability of urban spaces and their attractiveness for pedestrian use (Santucci, Fugiglando, Li, Auer, & Ratti, 2018). A higher variance of the SVF corresponds to a higher variability of the microclimatic conditions, producing frequent differences and variations in terms of outdoor comfort conditions: people preferably walk where the urban morphology determines variant microclimatic conditions. From a physiological point of view, sudden changes do not immediately provoke skin temperature shifts (Chokhachian, Santucci, Vohlidka, & Auer, 2017; Parkinson et al., 2012).

Impact

The concept of temporary appropriation has been widely discussed through past years as a key element of public space phenomena and in this regard Lara-Hernandez and Melis (2018) argued that there is a strong link between TA and social sustainability. What we tried to highlight here is the further relationship of TA to environmental criteria where the keywords like Health and wellbeing in cities have more room to be discussed and integrated into the equation of urban livability. The discussed methodology toward personalized comfort models can help us to first get better understanding of correlations between built environment in our cities and how people experience it in two different dimensions of psychology and physiology. This understanding of human centered urban fabric will enable designers and urban planners to design more walkable and likeable public spaces. This whole study is aligned with the hypothesis that microclimate is one of the primary factors that enables temporary (and not only) appropriation of public spaces which needs to be investigated more with different methodological tools. Moreover, the discussed approaches have significant implications for the understanding of how the design of the built environment could have the capability to allow the TA for future generations, which is one of the main aims of making cities more livable and enjoyable.

Acknowledgements The experiment in Singapore is the result of a cooperation between the Chinese University of Hong Kong, Institute of Future Cities, the University of California Berkeley, Center for the Built Environment and the Technical University of Munich, Chair of Building technology and Climate Responsive deign, funded by the TUM Global Incentive Fund. The experiment in Rome was funded by the DAAD—Deutscher Akademischer Austauschdienst, within a cooperation with La Sapienza University in Rome, Dipartimento di Pianificazione, design e tecnologia dell' Architettura, Prof. Alessandra Battisti and her students who contributed in visualizing the data.

References

Ahearne, J. (2010). Michel de Certeau. The practice of everyday life. *International Journal of Cultural Policy, 16*(1), 2–3. https://doi.org/10.1080/10286630902971595.

Cabanac, M. (1992). Pleasure: The common currency. *Journal of Theoretical Biology, 155*(2), 173–200. https://doi.org/10.1016/S0022-5193(05)80594-6.

Certeau, M. De. (1984). *The practice of every day life*. Berkeley, CA: University of California Press.

Chokhachian, A., Ka-Lun Lau, K., Perini, K., & Auer, T. (2018). Sensing transient outdoor comfort: A georeferenced method to monitor and map microclimate. *Journal of Building Engineering, 20*, 94–104. https://doi.org/10.1016/j.jobe.2018.07.003.

Chokhachian, A., Santucci, D., & Auer, T. (2017). A human-centered approach to enhance urban resilience, implications and application to improve outdoor comfort in dense urban spaces. *Buildings, 7*(4), 113.

Chokhachian, A., Santucci, D., Vohlidka, D.-I. P., & Auer, T. (2017). Framework for defining a transient outdoor comfort model in dense urban spaces, processes & findings. In Paper presented at the 7th International Doctoral Conference, Architecture and Urbanism: Contemporary Research.

De Dear, R., & Brager, G. S. (1998). *Developing an adaptive model of thermal comfort and preference*.

Haas, T., & Olsson, K. (2014). Transmutation and reinvention of public spaces through ideals of urban planning and design. *Space and Culture, 17*(1), 59–68. https://doi.org/10.1177/1206331213493855.

Hensen, J. L. M. (1990). Literature review on thermal comfort in transient conditions. *Building and Environment, 25*(4), 309–316. https://doi.org/10.1016/0360-1323(90)90004-B.

Johansson, E., & Emmanuel, R. (2006). The influence of urban design on outdoor thermal comfort in the hot, humid city of Colombo, Sri Lanka. *International Journal of Biometeorology, 51*(2), 119–133. https://doi.org/10.1007/s00484-006-0047-6.

Lara-Hernandez, J. A., & Melis, A. (2018). Understanding the temporary appropriation in relationship to social sustainability. *Sustainable Cities and Society, 39*, 366–374. https://doi.org/10.1016/j.scs.2018.03.004.

Lefebvre, H., & Nicholson-Smith, D. (1991). *The production of space*, Vol. 142. Oxford: Blackwell.

Liabäck Löwstett, F. (2018). *Enabling temporary use of public space* (Degree of Master). Sweden: KTH.

Parkinson, T., de Dear, R., & Candido, C. (2012). *Perception of transient thermal environments: Pleasure and alliesthesia*. In Paper Presented at the Proceedings of 7th Windsor Conference. UK: Windsor.

Roesler, S., & Kobi, M. (2018). Microclimates and the city, towards an architectural theory of thermal diversity. In S. Roesler & M. Kobi (Eds.), *The urban microclimate as artifact*. Basel: Birkhäuser.

Santucci, D., Fugiglando, U., Li, X., Auer, T., & Ratti, C. (2018). *Methodological framework for evaluating liveability of urban spaces through a human centred approach*. In Paper Presented at the 10th Windsor Conference Rethinking Comfort, NCEUB 2018, Windsor.

Speaks, M., Crawford, M., Mehrotra, R., & Kelbaugh, D. (2005). *Everyday urbanism*. University of Michigan, Taubman College of Architecture and Urban Planning.

Origins of Informality: Examining the Historical and Spatial Roots of Informal Day-Labor Hiring Sites

Sean M. Crotty

Abstract Informal day-labor hiring sites, public and semi-spaces where (mostly) men wait to be hired for short-term work, are found in cities throughout the United States. The daily-appropriation of space for labor-seeking is not a new phenomenon, however media and academic accounts of day-labor hiring sites almost exclusively focus on the contingent nature of laborers' employment to suggest that this use of space is socially unsustainable. Despite the fact that day-labor is a spatial means of employment-seeking, the process by which particular spaces are selected for day-labor hiring sites remains almost entirely unexamined. Drawing on more than five years of mixed-methods research in the San Diego Metropolitan Area, I examine the places where hiring sites operate to demonstrate the linkages between neighborhood characteristics and the establishment of day-labor hiring sites. Ethnographic data further supports an argument that hiring sites exhibit a high-degree of locational "stickiness," due to the historical, economic, and emotional content embedded in each site through the production of space. The paper concludes by examining the specific ways that laborers and other stakeholders interact to ensure that access to space is maintained and in so doing, foster relationships that produce socially-sustainable outcomes for day-laborers and their neighbors.

Keywords Day labor · Informal economy · Historical GIS · Social sustainability · Relational theory

Introduction

Life as a day-laborer in the San Diego Metropolitan Area (SDMA) is not easy. Every morning day-laborers, also known as jornaleros, travel to particular public and semi-public spaces where they wait for mostly short-term jobs: providing manual labor or general handyman services for homeowners, construction and landscaping companies, and a variety of other low-to-medium skill jobs. According to Abel Valenzuela, the term day-labor is generally used to "convey a type of temporary

S. M. Crotty (✉)
Texas Christian University, Fort Worth, USA
e-mail: sean.crotty@tcu.edu

A. Melis et al. (eds.), *Temporary Appropriation in Cities*,
https://doi.org/10.1007/978-3-030-32120-8_8

Fig. 1 A typical informal day-labour space in the SDMA

employment that is distinguished by hazards in or undesirability of the work, the absence of fringe and other typical workplace benefits (i.e., breaks, safety equipment), and the daily search for employment" (2000: p. 308). This definition does a nice job of highlighting the challenging, and often exploitive, nature of day labor work (Meléndez, Visser, Theodore, & Valenzuela, 2014). However, it fails to address a crucial element in the employment-search process. *Day labor is a form of employment seeking that depends on the temporary appropriation of public and semi-public spaces.*

The majority of laborers in the SDMA congregate at informally-designated hiring spaces on sidewalks and the edges of parking lots like the one shown in Fig. 1. Jornaleros typically arrive at the day-labour[1] space around 6:00am and will wait until 3:00–5:00pm in hopes of finding employment. Day-laborers do not own the spaces that they appropriate for their daily employment search, and yet, through a variety of individual and group actions, jornaleros maintain access to the spaces they believe provide their best chance for employment, safety, and survival. This is to say that day-labour spaces exhibit a high degree of locational stickiness. Once a hiring

[1]Crotty (2015) defines day-labour neighbourhoods as all census tracts located within ½ mile of a day-labour hiring site.

site is established, it tends to remain active in the same location over considerable periods of time—in San Diego many sites have be in consistent operation for 30 or more years.

In this chapter I situate day labour within literature on temporary appropriation (TA) and sustainability by examining the life-cycle of day-labour spaces. The field of urban sustainability is of critical importance as more than half the world's population currently resides in urban environments, and the percentage of urban inhabitants is projected to increase to 68% by 2050 (United Nations, 2018). Informal economic activity is also an increasingly important element of global economic activity (Danesh, 1991; Harding & Jenkins, 1989; Ihrig & Moe, 2004; Portes, Castells, & Benton, 1989). Neoliberal economic restructuring led to declines in formal-sector and organized labour arrangements while informal, contingent, and, most recently, "gig"-based economic arrangements became more prevalent (Aguilar & Herod, 2006; Castree, 2010; Hackworth & Moriah, 2006; Theodore, 2007). Many of these new arrangements depend on the temporary appropriation of urban space for success. Street vendors, unlicensed cabs and Uber drivers, and informal day-laborers all temporarily appropriate particular urban spaces as part of their income-generating strategies. Yet, the sustainability of these new urban economic arrangements remains relatively unstudied. Examining the processes through which particular spaces are appropriated for day labour activity, and examining the processes which lead laborers to abandon particular spaces, provides empirical context for improved understanding of the socio-spatial processes through which temporary appropriations of space occur. These processes speak directly to everyday power relations in the city and the sustainability of informal socio-economic activity in contemporary urban environments around the world. As such, this research contributes to work in urban planning and geography, landscape architecture, and other fields interested in TA and sustainability in the urban context.

Research Area and Methods

The analysis below draws on data collected from 2005 to 2012 as part of a mixed method research project in the San Diego Metropolitan Area (SDMA). The research included site-mapping, participant observation, and informal interview in or around day-labor hiring sites. Forty-five hiring sites were identified in total (See Fig. 2). After a site was identified it was surveyed at least four times per year at which time the number of laborers using the site and the demographic characteristics of those laborers was recorded (race, gender, age). Informal interviews were also conducted approximately once per year at each site to improve place-based understanding of day labor activities and changes in those daily activities. Finally, six sites were selected for in-depth participant observation based on particular site-characteristics found to be common among many SDMA hiring sites. These periods of participant observation

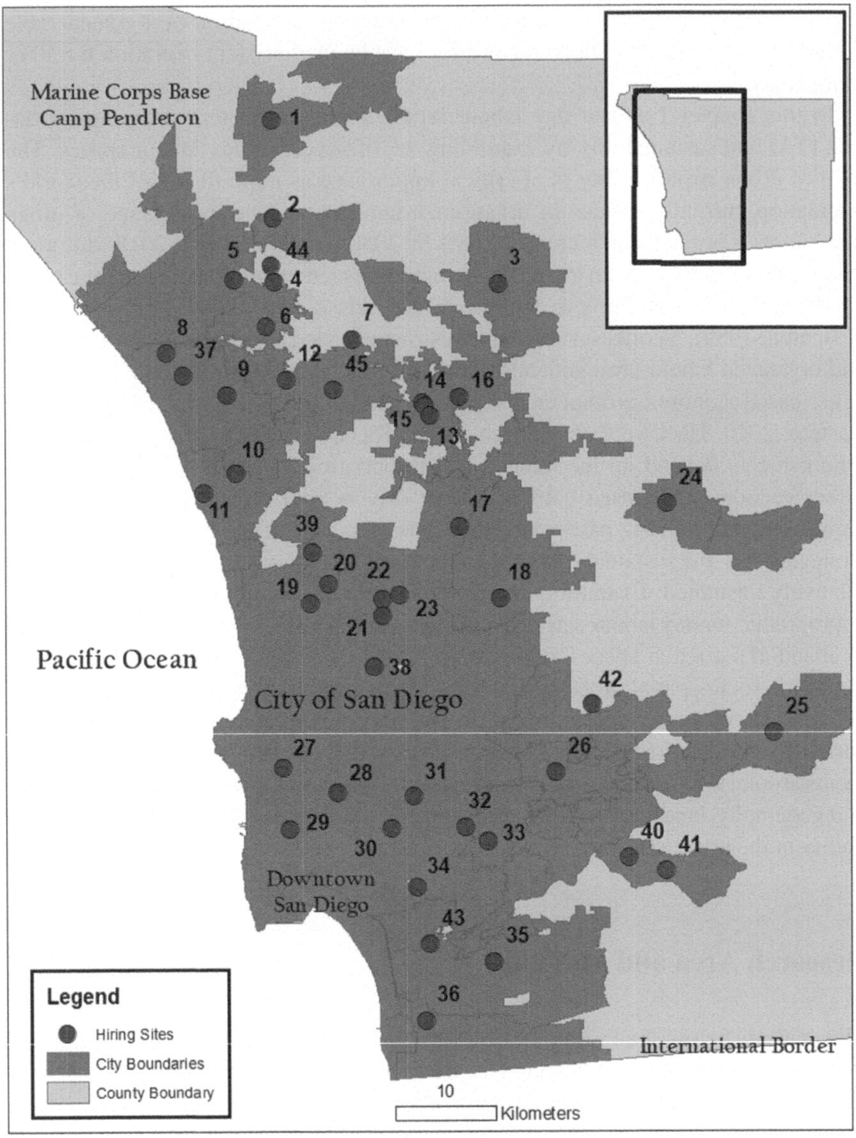

Fig. 2 Day-labour hiring sites in the San Diego metropolitan area

lasted between two and six weeks, during which time I visited the site every day for up to eight hours. This long-term qualitative engagement provides substantial data for situating day-laborer's appropriation of space in the context of social sustainability.

Day Labour and Social Sustainability

At first glance, day-labour may seem an unlikely candidate for inclusion in urban sustainability studies. Much of the existing research on the day-labor industry is focused on the challenging and often exploitive nature of the industry. Nonetheless, the long-term presence of day-labour hiring sites within the landscape of North American cities suggests that they are a quasi-permanent element of contemporary urban spaces and their sustainability should be evaluated in the same way that previous research has examined informal street markets (Yeo & Heng, 2014), urban plazas, gentrification processes (Lees, 2008), or urban design more broadly (Jacobs, 1961).

Sustainability studies are centred on three topical pillars: environmental, economic, and social sustainability. In the urban context sustainability these pillars frame diverse research focused on correctives for existing environmental, economic, and social injustices and/or inequities. The role of geography is critical in all of these cases as inequity is often materialized spatially. For example, toxic industrial emissions do not affect all residents equivalently. Those who live (or work) in close proximity to the emission site (or downwind, downriver, as the case may be) are disproportionately burdened compared to residents whose daily activity path never enters areas where exposure is high (Harwood, 2003; Openshaw, 1982; Pulido, 2000; Pulido, Sidawi, & Vos, 2005). Economic and social inequities also manifest spatially. The physical disconnection of low-income residential zones and clusters of employment opportunity in cities today is perhaps the most common example of this phenomenon in contemporary urban geography (McLafferty & Preston, 1992; Pastor Jr & Marcelli, 2000; Powell, 2002). The solutions to these spatial inequities include increased residential and commercial density, improvements to mass transit systems, and new urbanist or green architectural design, just to name a few. The sub-field of urban social sustainability has emerged in response to "the underlying tension between the associated aspects of sustainability—environmental, social, economic—as well as the wide interpretation of the concept" (Dempsey, Bramley, Power, & Brown, 2011: p. 289). Reviewing existing literature on urban social sustainability Dempsey et al. (2011) identify twenty non-physical factors related to social sustainability such as education and training, social capital, inclusion, social cohesion, and cultural traditions. By comparison, they identify eight factors that are predominantly physical in nature: urbanity, accessibility, pedestrian friendly-neighbourhoods, and an attractive public realm, are a few examples. The public realm is particularly important in the context of temporary appropriation and day-labour studies. Who has access to public spaces and how those spaces can (or should) be used is both a question of physical design and norms of community behaviour (Crotty, 2014; Mitchell, 2003; Staeheli

& Mitchell, 2008; Varsanyi, 2008). Yeo and Heng's definition of public space proves quite useful for merging existing research on public space and social sustainability. Their work "views public space as a resource with the potential to stimulate local economies, support household livelihoods, enhance civic engagement, promote community identity, and improve overall quality of life in communities" (Yeo & Heng, 2014, p. 715).

Previous day-labour research often emphasized the role of labourers as members of the local community through markers of their individual and collective social capital. For example, Valenzuela, Theodore, Melendez, and Gonzalez (2006) found that twenty-nine percent of *jornaleros* are raising children in the communities where they live and work. Many attend local churches while others are connected to community support organizations as well as organized labour groups (Fine, 2005, 2006; Meléndez et al., 2014; Visser, Theodore, Melendez, & Valenzuela, 2017). Taken together, these measures are evidence that labourers are not transient migrant labourers as anti-immigrant and anti-day-labour activists often suggest, but active members of the local community. This type of participation in community activities does more than just build social capital, it foments emotional connections to place which make labourers less-likely to alter their labour-seeking methods, and contribute to the social sustainability of temporary appropriations of space for day-labour.

Temporary Appropriation and Day-Labour Spaces

Appropriation of space is the process by which a particular space is used for a purpose other than its designers originally, or primarily, intended. Much of the research within the discipline of geography that deals with appropriation of space draws on the work of Lefebvre (1991) who defines appropriated space in relation to dominated space. Dominated space is "a space transformed – and mediated – by technology, by practice." (1991: p. 194). For Lefebvre, examples of dominated space include highways, dams, military infrastructure, and any other built elements that force their form onto natural environments in unnatural ways. The rectilinear grid of roads that forms the foundation of many urban and suburban environments would absolutely be considered dominated space. He further argues that dominated spaces are usually "closed, sterile, emptied out" (p. 165). Appropriated spaces are by contrast, lived spaces, where the rhythms of everyday life play out in ways that "resembles a work of art" (p. 165). Crucial in this distinction is that appropriated spaces are used in ways that maximize their *use-value* for urban residents irrespective of the designer's original intent. How, and by whom, spaces are appropriated "cannot be understood apart from the rhythms of time and life" (p. 166). In this way, Lefebvre's definition of appropriated space is temporarily mediated, and by necessity, temporary in nature.

Any examination of temporary appropriation of space is by extension an examination of power relations through the control of space. As John Allen notes, "if power has any trace at all, it is found *in place*." (Allen, 2003: p. 11, emphasis in original).

As such, the relative power of particular groups can be understood through the lens of temporary appropriation and control of space. And yet, not all spaces are equally important or useful to all individuals or groups. Therefore a real understanding of power relations requires understanding the objectives of particular individual and/or groups engaged in temporary appropriation of space and evaluating how closely their desires match their use of particular urban spaces. For example, in an examination of migrant minority Filipinos' religious processions in Padua, Italy, Chantal Saint Blancat and Adriano Cancellieri find that the Filipino migrants seek out high-traffic, highly-visible public spaces for their rituals. The religious processions accomplish two important objectives for the Filipino residents of Padua. First, it helps them maintain personal and group identity by continuing cultural practices while they are living abroad. The second objective is to improve acceptance and understanding of the Filipino community among the native Italian residents of Padua. That second objective establishes spatial requirements for the processions that would not otherwise exist. Improved understanding and acceptance require that as many Italians as possible are exposed to the Filipino cultural activities. Places with high pedestrian traffic and high visibility therefore have greater use-value than do more private, less-visible urban spaces (Saint-Blancat & Cancellieri, 2014). Conversely, places with low-visibility and minimal pedestrian traffic would have higher use-value for homeless individuals seeking a place to rest undisturbed for the evening (Mitchell, 2003). Staeheli and Mitchell's examination of community gardens in New York City demonstrates the role of accessibility, as well as environmental factors such as hours of sunlight and south-facing orientation, in producing spatial variation of use-value for prospective garden locations (2008).

Temporary appropriation for more explicitly economic purposes can produce yet another topography of use-value. For informal street vendors in Peru, sidewalks and under-utilized public or semi-public spaces along heavily-used pedestrian travel routes are valuable during commuting hours in the morning and afternoon, while similar spaces near central business districts or plazas have a higher use-value during working hours and weekends. These vendors, also known as *ambulantes*, negotiate access to their preferred spaces with other stakeholders, including formal-sector shopkeepers who establish relationships of mutual benefit with the *ambulantes* (Hays-Mitchell, 1994, 1999, 2002). This strategic use of space can even be observed in seasonal fundraising cookies sales by members of the Girl Scouts of America. Girls Scouts set up tables to sell cookies outside of grocery stores, shopping malls, sporting events, and other high traffic areas (Goerisch and Swanson 2013; Goerisch 2018). The girl scouts also understand that particular groups of people are more likely to purchase cookies than others and have responded by establishing cookie sales outside of marijuana dispensaries in San Diego (Stump 2018), Los Angeles (Harris, 2014), Colorado (Allen, 2019), and Edmonton, Ontario (Ross, 2018). Fewer total customers visit marijuana dispensaries than a shopping mall or grocery store, however the likelihood of dispensary customer purchasing cookies is apparently considerably higher than is the case for the general public, and as a result, the spaces outside of marijuana dispensaries have a high use value for Girl Scouts during cookie sales season.

Places are in a constant state of (re)production and are shaped by processes operating across multiple scales. This process of (re)production can change the nature of spaces and alter their "value" or "utility" for particular activities (Massey, 1991, 2005). These changes can include anything from seasonal weather patterns, to improved or worsening access to the space (either through mobility expansion or constraints), or changes in state policy/policing regarding particular spaces or actors. This is certainly true in the above examples and for day labour spaces as well.

Day-Labour Spaces

Day-labour researchers use a variety of terms to refer to the locations where day-labourers congregate each morning in hopes of finding employment. Hiring sites, shape-up sites, *la esquina*, and *la parada* are all used somewhat interchangeably and this inconsistency in terminology can occasionally lead to confusion regarding the place being analyzed (Ordóñez, 2015; Turnovsky, 2006; Valenzuela et al., 2006; Valenzuela, 2003). To address this issue, I refer to these locations as day-labour spaces and follow the definition established in Crotty (2018).

> Day-labour spaces are activity areas where *jornaleros* wait for potential employers, negotiate the terms of each job, eat, drink, socialize, and otherwise spend the time in between each employment negotiation. To be clear, day-labour spaces are not the locations where *jornaleros* perform paid labour; day-labour spaces are the general areas where *jornaleros* wait to be hired. Various actors produce day-labour spaces through informal negotiations that define where, when, and how day-labourers solicit employment. As a result of the constant negotiation and renegotiation between actors, sociospatial relationships are coproduced, and the boundaries of day-labour spaces become flexible and fluid. In this way, day-labour spaces are permeable, adaptable, and can be noncontiguous in some cases. (Crotty, 2018: p. 598)

The process by which day-labour spaces are established remains unstudied despite passages like the one below from noted economic geographer Nik Theodore. "The location of these informal hiring sites (*the designation of which, in many instances, occurs mysteriously*) varies depending on local circumstances: In Los Angeles, day labourers from Mexico and Central America gather at home improvement and building supply stores; in the municipality of South Tucson, Mexican workers assemble in the parking lot of a church that was a stalwart of the sanctuary movement; in Miami, workers from Haiti and the Dominican Republic congregate under highway overpasses; in Seattle, workers from Mexico and Central America take up positions on street corners just outside the central business district; In Chicago, workers from Eastern Europe gather at a Shell gas station; in Atlanta, workers from Mexico gather in the parking lot of an apartment complex; and in New York, Sikhs congregate near a burned-down mosque in Queens" (Theodore, 2007: pp. 257–258, emphasis added).

Theodore's description of site designation as mysterious and dependent on local circumstances suggests a gap in existing day-labour research that is critical to understanding the processes by which urban space is appropriated for informal economic uses. In all day-labour research, including work done by Geographers, there is little attention to the spatiality of day-labour at the regional or neighbourhood scale. This is fine in work like Theodore's, which focuses on structural processes at the national and global scale that increase contingent and temporary work like day-labour, but is less acceptable for researchers concerned about the day-to-day operation of the informal economy, temporary appropriation of space, sustainability of day-labour, or the locational conflicts that sometimes occur around day-labour hiring sites. Despite the lack of attention to spatial organization in much of the existing day-labour research, existing literature often contains anecdotal reference to site-locations that help to understand the temporary appropriation process. For example, in the early 1900s day-labour spaces existed in low-income "skid row" and dock-adjacent areas in US cities (Larrowe, 1955; Mohl, 1971; Valenzuela, 2003). In parts of the US where agriculture was the primary economic activity, workers congregated in town-centres to make themselves available to farmers who travelled into town from the periphery to hire short-term help (Mitchell, 1996).

This all changes with the suburbanization of US cities, for which the San Diego Metropolitan Area is a prime example. Figure 3 illustrates the extent of (sub)urban growth in the SDMA from 1950 to 2010. The historic urban boundaries are represented by the US Census tracts for San Diego in 1950 while the contemporary boundaries are comprised of the municipal borders for the 19 cities that collectively make up the San Diego Metropolitan Area today. The SDMA today covers more than double the total land area than was the case in 1950. Much of the land converted to suburban residential and commercial land use today was formerly used for agriculture that depended on day-labour to meet seasonal labour needs. Day-labour hiring sites would have been established in agricultural town centres and other spaces that made labourers accessible to the farmers that were their primary employers. These hiring sites may not have operated year-round as labour demand would fluctuate seasonally, even in San Diego's relatively stable climate. In the contemporary day-labour market, sites in the SDMA operate year-round, with some seasonal variation in the number of labourers congregating at particular sites. Though it is impossible to say with certainty when the shift from seasonal appropriation of space to year-round appropriation of space for day-labour activities takes place, the change is almost certainly linked to the shift from agricultural land use to suburban developments and the year-round labour required to build, and then maintain, the new suburban landscape.

The locations of day-labour hiring sites at the regional scale reflect the geography of employment demand. In an analysis of the demographic characteristics of day-labour neighbourhoods, Crotty (2015) found that day-labour neighbourhoods have higher levels of formal-sector employment in the industries in which labourers are most commonly hired: construction and agriculture. Homeowners who need short-term help with home-improvement projects make up the other half of day-labour employment (Valenzuela et al., 2006), and jornaleros clearly understand this as day-labour neighbourhoods also have higher levels of owner occupied housing compared with regional averages (Crotty, 2015). This clear regional strategy that targets areas of employment demand does not mean that the sites are homogeneous. Indeed a closer examination of the demographic characteristics of day-labour neighbourhoods and geographic site and situational characteristics of hiring sites reveals "types" of sites

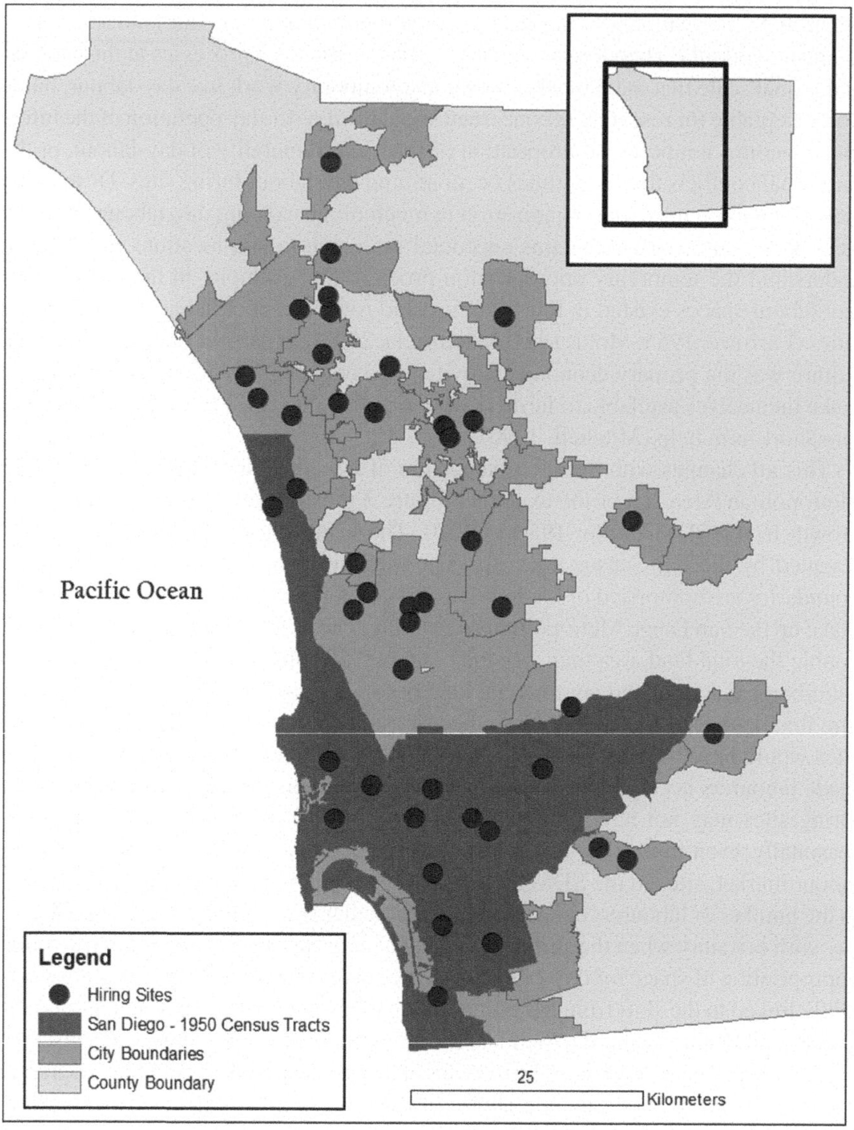

Fig. 3 San Diego metropolitan area: urban boundaries in 1950 and 2010

in the SDMA. And yet, simply acknowledging current characteristics of sites does not explain how or why a particular space was initially appropriated for day-labour hiring.

Case Study 1: Establishment of New Day-Labour Spaces

In the coastal municipality of Encinitas and Oceanside there are several day-labour hiring sites. According to local residents, the longest-operating site is site #11, located near the intersection of an interstate highway and the largest east-west road in the area. Site 11 was established sometime in the late 1960s, after the highway construction was completed. Thanks to its location near the intersection of two major mobility corridors, this site was easily accessible to potential employers across a wide physical area. At the time, much of the land to the west of the site was already developed as urban beach communities, while the land east of I-5 was largely agricultural. Agricultural work was the primary type of employment at site 11 from the late 1960s until at least the late 1980s. Figure 4 shows the change in agricultural land use from 1986 to 2008. During that 22 year span the SDMA saw a 65% decrease in agricultural land use, much of which was transitioned from agriculture to suburban land use. That shift also opened the market for home improvement and construction supply stores like the Home Depot location that opened in 1993.

The new store is located roughly three kilometres northeast of site 11. Shortly after the store opened, day-labourers began congregating in adjacent public and semi-public spaces. The establishment of the new site (#10) suggests that some labourers recognized that the space outside of Home Depot had a use-value that was at least as good as site 11. The position adjacent to a home improvement store makes *jornaleros* easily accessible for potential employers who need assistance with construction and home improvement projects. Home Depot is also considerably more popular with formal-sector construction workers than its largest competitor Lowes. Construction contractors are the main source of long-term employment for *jornaleros*, and because long-term employment is the goal for most labourers, they congregate in places where they believe their opportunity for long-term employment is greatest. Indeed, *jornaleros* seem to know at which particular Home Depot locations construction contractors comprise a larger percentage of the customer base, as day labour spaces are only active near 9 of the 21 Home Depot locations in the SDMA.

Despite the establishment of a new day-labour space just a few kilometres away, site 11 remains active today with roughly 20 *jornaleros* seeking work there daily. The labourers who seek work at site 11 are relatively diverse and even organize spatially along lines of nationality and legal status (Crotty, 2018). These labourers continue to seek work at site 11 for a variety of reasons. Some prefer the less-competitive nature of the site, others suggest there is less daily complaints and hassles from nearby stakeholders, and some claim that the jobs hired from the site are of better quality than those hired from the Home Depot. For these men, site 11 provides enough work to survive which, in combination with their individual preferences for type of

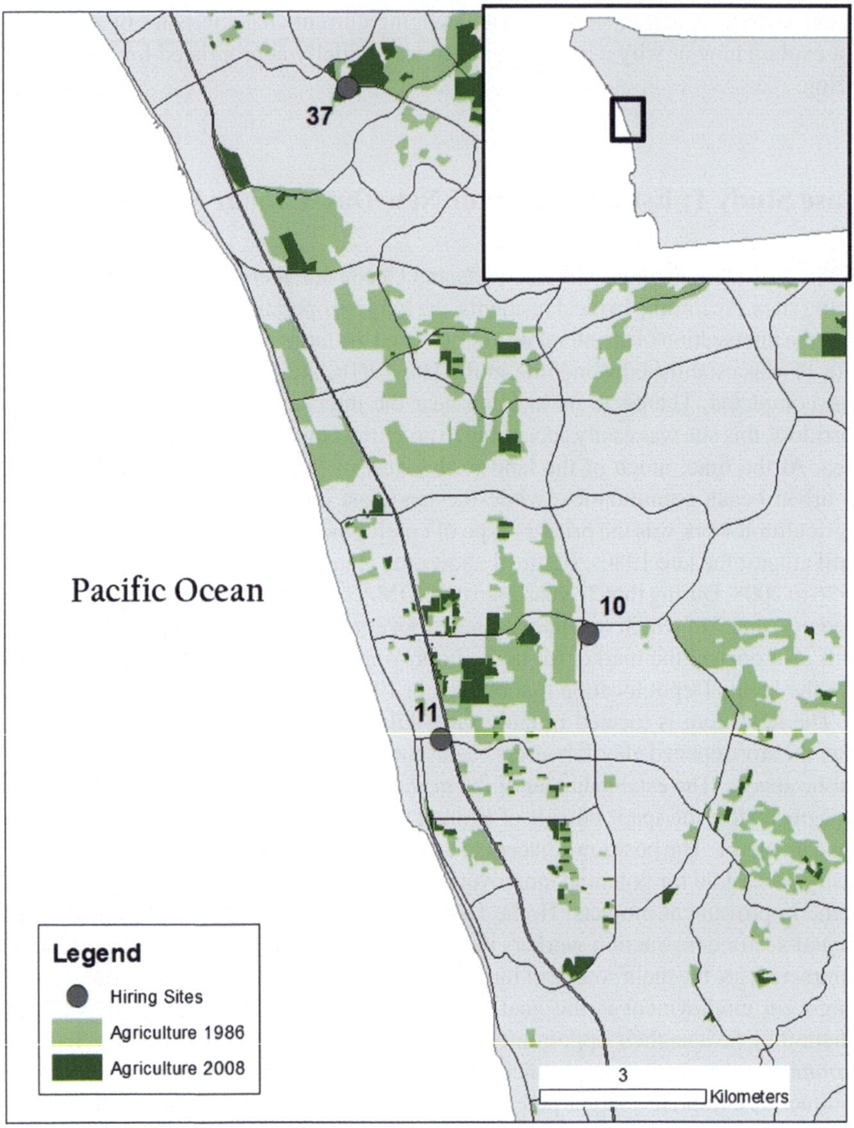

Fig. 4 Change in agricultural land use in Northwest San Diego County: 1986–2008

employment and site-specific behavioural norms, produces a higher use-value than the Home Depot adjacent location.

Sites 10 and 11 were both established long before this research project began. It is only through interviews with residents and archival research that it is possible to understand how and why these particular spaces were appropriated for day-labour activities. That is not the case for site 45, located adjacent to a Home Depot store in San Marcos, CA. Site 45 was established in 2008, when the local construction industry was decimated by the Great Recession. Initially site 45 was used almost exclusively by White and Black labourers for whom this HD was more easily accessible than others with established DL sites. Some of these new DLs also may not have wanted to compete with more-experienced Hispanic *jornaleros* at other sites, but there is little question that residential segregation patterns made this site more easily accessible for non-Hispanic labourers. By 2010 the labourers most of the labourers using the site were Hispanic, which is more reflective of the demographic trends for the region, but also may reflect the recovery of formal-sector construction employment. Those waylaid formal sector construction workers who established the day-labour space as a means of generating income to survive the downturn likely returned to full time employment as industry conditions improved. The movement of Hispanic labourers to this site due in part to anti-day-labour actions taken at a site in the adjacent municipality of Vista (See Crotty 2017 for more details about that anti-day-labour action). And yet, more than 10 years later the day-labour space remains in use by Hispanic labourers also looking for full-time work in the construction industry.

In each of the above cases, space was appropriated that was accessible to both employers and labourers. The spaces appropriated were not necessarily the most visible, but provided a high degree of utility for the labourers. In both cases, we see large-scale processes changing the nature of day-labour places, and labourers themselves adapting spatially by establishing new day-labour spaces in Encinitas: to meet demand for new types labour and to maintain a lower use-density in area day-labour spaces, and in San Marcos: initially, to provide newly unemployed labourers a place they felt safe seeking part-time work, and later, a space for more experienced labourers who did not want to be hassled by sheriffs and code-compliance officers in Vista. That all of these day-labour spaces remained active at the end of the research period, despite some private and state-sponsored efforts to relocate day-labour activity speaks to the power of everyday actions and temporary appropriation.

Labourers work hard to ensure continued access to the spaces they believe have the highest use-value for employment seeking. Crotty attributes *jornaleros* success in maintaining access to their preferred hiring sites and the resulting locational stickiness of day-labour spaces to "labourers' collective employment of *strategic visibility* and the production of normalcy through sustained presence of day-labour hiring sites in particular locations" (Crotty, 2018: p. 601). Strategic visibility is a set of informal spatial practices employed at the micro-geographic scale that reduce the potential for locational conflict while maintaining accessibility for potential employers. Labourers congregating on a low-traffic street near a major road, rather than on the major road itself would be an example of spatial practice common across day-labour spaces in the SDMA. From that position, labourers are conveniently accessible for potential employers, but reduce the potential for conflict because their hiring negotiations take place in a low-traffic space where their visibility and potential to create traffic hazards is minimized. The fact that *jornaleros* show up to the same places every day, in some cases for 30 years or more, produces a sense of normalcy within the rhythms of the city that further cements their claims to space (from 6am to 3pm daily).

Case Study 2: End of Appropriation; Death of a Day-Labour Space

In the context of day-labour spaces that operate in the same locations for decades at a time, the death of a site, when labourers cease to use the space entirely, is extraordinarily rare. The single instance of this occurring in the SDMA during the research period occurred at site 32 (See Fig. 5). Site 32 was never an especially large hiring site. In all of the field surveys of the site there were never more than 15 labourers there. It was notable within the research project because it was one of the first sites identified.

In 2006 there were several construction supply and tool rental shops in operation along that block of University Drive. Over the course of the research project, they all closed or relocated so there are currently no nearby businesses that serve the industries for which labourers are typically hired. This reduction in employment opportunities was due, in part, to expanded tool-rental services available at the Lemon Grove Home Depot, located just two kilometres to the southwest. Hiring site 33 operates adjacent to the Lemon Grove Home Depot and, in much the same way that competition from Home Depot reduced the number of customers frequenting the building supply and tool rental businesses along University Ave, competition from site 33 reduced the number of potential employers visiting site 32. Clearly economic competition played a role in the death of site 32, but it is not the only relevant consideration. Geographic site and situational characteristics of both day-labour spaces played a role in the demise of site 32 as well. Site 33 enjoys a number of geographic advantages relative to site 32: it is located directly adjacent to a stop on the regional light-rail system, so it is accessible to labourers who depend on public transit to reach their preferred hiring site. Site 33 is also located less than 100 meters from the intersection of two major highways, making it highly accessible for potential employers. In the end, site 32's use-value fell far enough that labourers could not justify continuing to appropriate the space for their daily search for employment.

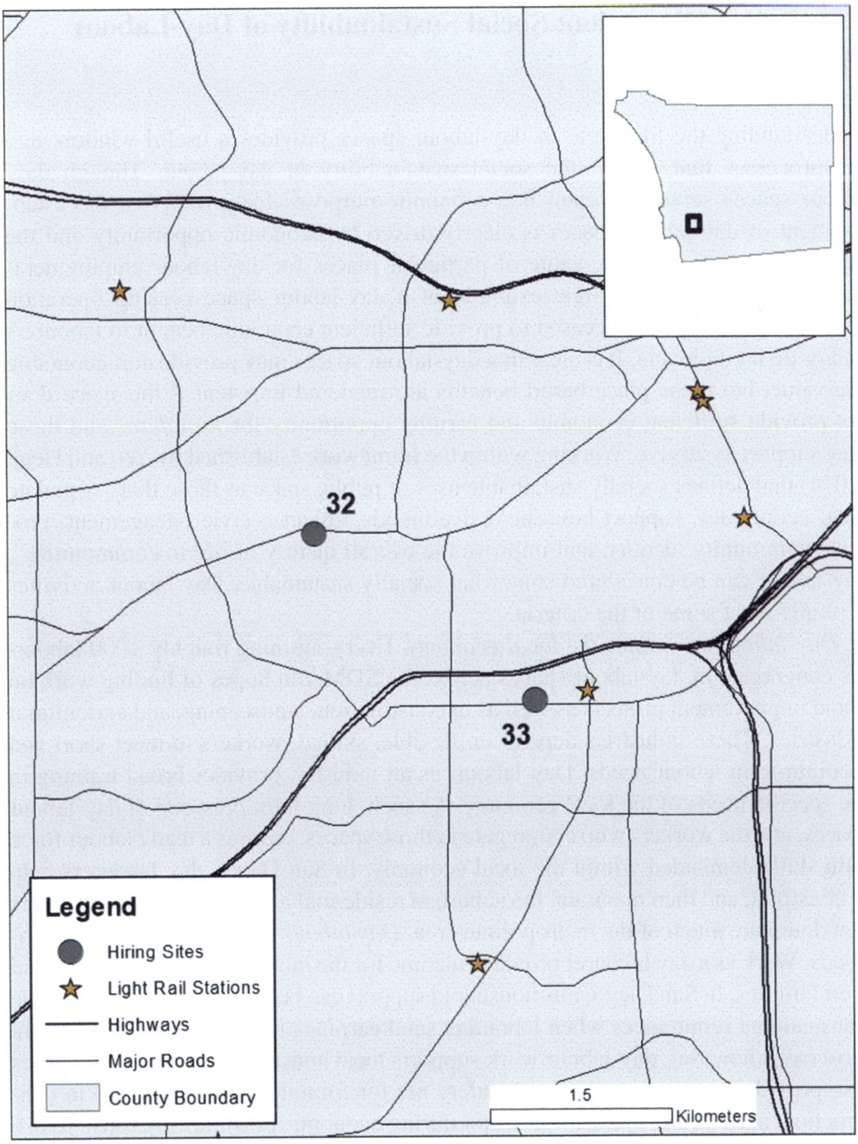

Fig. 5 Hiring sites in east San Diego

Concluding Discussion: Social Sustainability of Day-Labour Spaces

Understanding the life-cycle of day-labour spaces provides a useful window into the processes that impact the social sustainability of day-labour. Though day-labour spaces serve important non-economic purposes for *jornaleros*, the establishment of day-labour spaces is clearly driven by economic opportunity and the perceived/experienced use-value of particular places for day-labour employment-seeking. Similarly, the single example of a day-labour space ceasing operation occurred when the space ceased to provide sufficient economic benefit to labourers and/or their employers. It is clear that day-labour spaces may provide non-economic use-value, but those place-based benefits are rendered impotent if the space does not provide sufficient economic and earning opportunity for *jornaleros* and those they support to survive. Working within the framework established by Yeo and Heng (2014) that defines socially sustainable uses of public space as those that "stimulate local economies, support household livelihoods, enhance civic engagement, promote community identity, and improve the overall quality of life in communities", day-labour can be considered somewhat socially sustainable. Day labour activities certainly meet some of the criteria.

Day-labour stimulates the local economy. Every morning roughly 1000 labourers congregate in day-labour spaces across the SDMA in hopes of finding work on home improvement projects as well as in construction, landscaping, and agricultural industries. These industries depend on flexible, skilled, workers to meet short and medium-term labour needs. Day-labour, as an industry, provides broad training in the specific needs of the local economy. As such, long-term presence of day-labour spaces, and the workers who congregate in those spaces, ensures a ready labour force with skills demanded within the local economy. In San Diego, day labourers help to construct, and then maintain, the suburban residential and commercial landscapes that dominate much of the metropolitan area. *Day labour supports household livelihoods.* Work as a day labourer provides income for the more than 1000 labourers and their families. In San Diego, this household support can be local, but can also include transnational remittances when labourers send earnings to family living abroad. In most cases however, day-labour work supports local households. Day-labour spaces also provide a precarious, but real, safety net for formal-sector employees in construction, agriculture, and service sector during economic downturns (Crotty, 2014).

Beyond these two criteria, the social sustainability of day-labour spaces becomes less clear. *Day-labour spaces enhance civic engagement.* There is no question that the presence of day-labour spaces can increase civic engagement, but as the public-face of immigration-related conflicts in the United States, at least some of the 'engagement' day-labour spaces inspire is anti-social and may divide a geographically-defined community more than unite it (Esbenshade, 2000). Conversely, neighbourhood and/or municipal-level conflicts regarding the appropriation of space for day-labour can connect *jornaleros* to a community of support as well. Local faith-based organizations, immigrant-support organizations, and workers-rights advocates often mobilize to support day-labourers during periods of conflict. In this way, *day-labour conflict can promote community identity*. Conflicts about day-labourers appropriation of space are often more about racialized neighbourhood identity than any real-world challenges posed by the activities which occur in day-labour spaces (Esbenshade, 2000; Varsanyi, 2008). In this way, members of a community reveal their imagined identity through their collective response to day-labour spaces and through the types of day-labour management policies they support. In such conflicts, the concept of "quality of life" is mobilized by groups both in support of, and opposed to, day-labourers appropriation of space for their livelihoods. That both sides of a conflict can use the same concept in support of their position speaks to the subjective and nebulous way that quality of life is defined. As such, it is impossible to say with certainty whether or not *day-labour spaces promote overall quality of life*. How day-labour spaces affect other residents quality of life is unclear. As the occasional locational-conflicts around day-labour spaces make clear, some residents feel threatened by groups of men congregating in public spaces and thereby experience a reduced quality of life. Conversely, some residents hire *jornaleros* to provide low-cost, high-quality work for home-maintenance projects, yard care, even specialized tasks like plumbing, roofing, or masonry. These residents quality of life is thereby improved. The third, and largest, group of residents are those who do not view day-labour spaces as threatening, but also do not directly hire *jornaleros* themselves. For this group, it is hard to argue that day-labour has any measurable impact on their quality of life. Thanks to *jornaleros* collective strategic visibility and the resulting locational stickiness of day-labour spaces, hiring site become just another element in the landscape for residents going about their daily routines. Whether day-labour spaces promote the quality of life for *jornaleros* is also somewhat unclear. Day-labour life is challenging. Workers-rights violations are common. Poverty is cyclical under the best circumstances and endemic to the industry in most circumstances (Crotty, 2014; Valenzuela et al., 2006). Yet, *jornaleros* survive. They show up every morning in hopes of better employment. And collectively, they produce spaces that help sustain them socially, economically, and even spiritually (Crotty, 2014, 2018; Sifuentes, 2007). Day-labourers' income and social outcomes compare favorably to jobs in grocery, restaurant, or janitorial services they might take instead (Doussard, 2013). Day-labour work is generally inclusive. Despite the competitive nature of their contingent employment arrangements, place-based norms of behavior help new arrivals to each hiring site learn how to best find a job for the day. Over time social bonds between labourers, forged within day-labour spaces, lead to a host of actions that ensure group survival over individual financial success (Crotty, 2014, 2018). And what is sustainability about if not group survival?

References

Aguilar, L. L. M., & Herod, A. (2006). *The dirty work of neoliberalism: cleaners in the global economy*. John Wiley & Sons.

Allen, E. (2019). Girls Scouts in Colorado given green light to sell cookies outside pot shops|FOX31 Denver. [Online]. Colorado's own news. Available from https://kdvr.com/2019/02/05/girls-scouts-in-colorado-given-green-light-to-sell-cookies-outside-pot-shops/. Accessed March 26, 2019.

Allen, J. (2003). *Lost geographies of power*. Malden: Blackwell Publishers Ltd.

Castree, N. (2010). Crisis, continuity and change: Neoliberalism, the left and the future of capitalism. *Antipode, 41*(S1), 185–213.

Crotty, S. M. (2018). Strategic visibility and the production of day-labor spaces: A case study from the San Diego metropolitan area. *Urban Affairs Review, 54*(3), 593–631.

Crotty, S. M. (2014). The social geography of day labor: Informal responses to the economic downturn. *Yearbook of the Association of Pacific Coast Geographers, 76,* 22–48.

Crotty, S. M. (2015). Locating day-labor employment: Toward a geographic understanding of day-labor hiring site locations in the San Diego metropolitan area. *Urban Geography, 36*(7), 993–1017.

Crotty, S. M. (2017). Can the informal economy be "Managed"?: Comparing approaches and effectiveness of day-labor management policies in the San Diego Metropolitan Area. *Growth and Change, 48*(4), 909–941.

Danesh, A. H. (1991). *The informal economy: Underground economy, moonlighting, subcontracting, household economy, unorganized sector, barter, ghetto economy, second economy; a research guide*. Garland.

Dempsey, N., Bramley, G., Power, S., & Brown, C. (2011). The social dimension of sustainable development: Defining urban social sustainability. *Sustainable Development*. [Online], 289–300. Available from https://doi.org/10.1002/sd.417. May 19, 2009.

Doussard, M. (2013). *Degraded work: The struggle at the bottom of the labor market*. U of Minnesota Press.

Esbenshade, J. (2000). The "crisis" over day labor: The politics of visibility and public space. *WorkingUSA, 3*(6), 27–70.

Fine, J. (2005). Community unions and the revival of the american labor movement. *Politics Society*, [Online] *33*(1), 153–199. Available from https://doi.org/10.1177/0032329204272553.

Fine, J. (2006). *Worker centers: Organizing communities at the edge of the dream*. Ithaca: ILR Press/Cornell University Press.

Goerisch, D. (2018). Operation thin mint: Popular geopolitics of care and post-9/11 girlhood. *Young, 27*(2), 108–122.

Goerisch, D., & Swanson, K. (2013). 'It's called Girl Scouts, not, like, Woman Scouts': Emotional labour and girls' bodies. *Children's Geographies, 13*(4), 451–466.

Hackworth, J., & Moriah, A. (2006). Neoliberalism, contingency and urban policy: The case of social housing in Ontario. *International Journal of Urban and Regional Research, 30,* 510–527.

Harding, P., & Jenkins, R. (1989). *The myth of the hidden economy: Towards a new understanding of informal economic activity, 7*(4), 739–741.

Harris, J. (2014). Girl Scout sells cookies outside pot dispensary: 117 boxes in 2 hours. Los Angeles Times. [Online] February 14. Available from https://www.latimes.com/food/dailydish/la-dd-girl-scout-sells-cookies-pot-clinic-20140221-story.html.

Harwood, S. A. (2003). Environmental justice on the streets: Advocacy planning as a tool to contest environmental racism. *Journal of Planning Education and Research*, [Online] *23*(1), 24–38. Available from https://doi.org/10.1177/0739456x03255431.

Hays-Mitchell, M. (1999). From survivor to entrepreneur: Gendered dimensions of microenterprise development in Peru. *Environment and Planning A*, [Online] *31*(2), 251–271. Available from https://doi.org/10.1068/a310251.

Hays-Mitchell, M. (2002). Resisting austerity: A gendered perspective on neo-liberal restructing in Peru. *Gender and Development*, [Online] *10*(3), 71–81. Available from https://doi.org/10.1080/13552070215920.

Hays-Mitchell, M. (1994). Streetvending in Peruvian cities: The spatio-temporal behavior of ambulantes. *The Professional Geographer*, [Online] *46*(4), 425–438. Available from https://doi.org/10.1111/j.0033-0124.1994.00425.x.

Ihrig, J., & Moe, K. S. (2004). Lurking in the shadows: the informal sector and government policy. *Journal of Development Economics, 73*(2), 541–557.

Jacobs, J. (1961). *The death and life of great American cities*. New York: Random House.

Larrowe, C. P. (1955). *Shape-up and hiring hall: A comparison of hiring methods and labor relations on the New York and Seattle water fronts*. University of California Press.

Lees, L. (2008). Gentrification and social mixing: Towards an inclusive urban renaissance? *Urban Studies, 45*(12), 2449–2470.

Lefebvre, H. (1991). *The production of space* (Vol. 142). (Nicholson-Smith, D., Trans.). Malden: Blackwell Publishing.

Massey, D. (1991). A global sense of place. Marxism today. *Health and Inequality, 308*(June), 24–28.

Massey, D. (2005). *For Space*. [Online]. Wilshire, Sage Publications Ltd. Available from https://selforganizedseminar.files.wordpress.com/2011/07/massey-for_space.pdf.

McLafferty, S., & Preston, V. (1992). Spatial mismatch and labor market segmentation for African-American and Latina women. *Conomic Geography*, *68*(4), 406–431.

Meléndez, E. J., Visser, M. A., Theodore, N., & Valenzuela, A., Jr. (2014). Worker centers and day laborers' wages. *Social Science Quarterly, 95*(3), 835–851.

Mitchell, D. (2003). The end of public Space? The right to the city—Social justice and the fight for public space. [Online]. 5608 May 2014, 118–160. Available from https://doi.org/10.1111/j.1467-8306.1995.tb01797.xa.

Mitchell, D. (1996). *The lie of the land: Migrant workers and the California landscape*. U of Minnesota Press.

Mohl, R. A. (1971). *Poverty in New York, 1783–1825*. USA: Oxford University Press.

Openshaw, S. (1982). The geography of reactor siting policies in the U.K. *Transactions of the Institute of British Geographers*, *7*(2), 150–162.

Ordóñez, J. T. (2015). *Jornalero: Being a day laborer in the USA*. Los Angeles: California Series in Public Anthropology.

Pastor Jr, M., & Marcelli, E. A. (2000). Men in the hood: Skil, spatial, and social mismatch among workers in Los Angeles County. *Urban Geography*, [Online] *21*(6), 474–496. Available from https://doi.org/10.2747/0272-3638.21.6.474.

Portes, A., Castells, M., & Benton, L. A. (1989). *The informal economy: Studies in advanced and less developed countries*. JHU Press.

Powell, J. (2002). Prawl, fragmentation, and the persistence of racial inequality: Limiting civil rights by fragmenting space. In G. D. Squires (Ed.), *Urban Sprawl: Causes, consequences, and policy responses* (pp. 73–117). The Urban Institute: Washington D.C.

Pulido, L. (2000). Rethinking environmental racism: White privilege and urban development in Southern California. *Annals of the Association of American Geographers*, *90*(1).

Pulido, L., Sidawi, S., & Vos, R. O. (2005). An archeology of environmental racism in Los Angeles. In N. R. Fyfe & J. T. Kenny (Eds.), *The urban geography reader* (pp. 229–240). New York: Routledge.

Ross, A. (2018). Edmonton girl guide sells out of cookies in cannabis store lineup|CBC News. [Online] October 18. Available from https://www.cbc.ca/news/canada/edmonton/girl-guide-cookies-cannabis-lineup-1.4868520. Accessed March 26, 2019.

Saint-Blancat, C., & Cancellieri, A. (2014). From invisibility to visibility? The appropriation of public space through a religious ritual: The Filipino procession of Santacruzan in Padua, Italy. *Social & Cultural Geography*. [Online], *15*(6), 645–663. Available from https://doi.org/10.1080/14649365.2013.879494.

Sifuentes, E. (2007). *Documentary focuses on migrant 'chapel': Local filmmaker tells story of its demolition*. San Diego Union Tribune. [Online] July 11. Available from https://www.sandiegouniontribune.com/sdut-documentary-focuses-on-migrant-chapel-local-2007jul11-story.html.

Staeheli, L. A., & Mitchell, D. (2008). *The people's property?: power, politics, and the public*. New York: Routledge.

Stump, S. (2018). Girl Scout sells 300 boxes of cookies outside of marijuana dispensary in 6 hours. Today.com. Retrieved from https://www.today.com/parents/girl-scout-sells-300-boxes-cookies-outside-marijuana-dispensary-6-t122455.

Theodore, N. (2007). Closed Borders, Open Markets: Immigrant Day Laborers' Struggle for Economic Rights. In H. Leitner, J. Peck, & E. S. Sheppard (Eds.), *Contesting neoliberalism: Urban frontiers* (pp. 250–265). New York: The Gildford Press.

Turnovsky, C. P. (2006). A la parada: The social practices of men on a street corner. *Social Text*, [Online] *24*(3(88)), 55–72. Available from https://doi.org/10.1215/01642472-2006-004.

United Nations. (2018). 68% of the world population projected to live in urban areas by 2050, says UN| UN DESA|United Nations Department of Economic and Social Affairs. [Online] Available from https://www.un.org/development/desa/en/news/population/2018-revision-of-world-urbanization-prospects.html. Accessed: September 17, 2018.

Valenzuela, A. (2003). Day labor work. *Annual Review of Sociology*, [Online] *29*(1), 307–333. Available from https://doi.org/10.1146/annurev.soc.29.010202.100044.

Valenzuela, J. A., Theodore, N., Melendez, E., & Gonzalez, A. L. (2006). On the corner: Day labor in the united states. [Online]. Available from http://www.sscnet.ucla.edu/issr/csup/index.php.

Valenzuela Jr, A. (2000) Working on the margins: Immigrant day labor characteristics and prospects for employment. *The Center for Comparison Inmmigration Studies*, (22).

Varsanyi, M. W. (2008). Immigration policing through the backdoor: City Ordinances, the 'Right to the City', and the exclusion of undocumented day laborers. *Urban Geography.*, *29*(1), 29–52.

Visser, M. A., Theodore, N., Melendez, E. J., & Valenzuela, A. (2017). From economic integration to socioeconomic inclusion: Day labor worker centers as social intermediaries. *Urban Geography*, [Online] *38*(2), 243–265. Available from https://doi.org/10.1080/02723638.2016.1168574.

Yeo, S.-J., & Heng, C. K. (2014). An (extra) ordinary night out: Urban informality, social sustainability and the night-time economy. *Urban Studies, 51*(4), 712–726.

Unsheltered Homelessness and the Right to Metabolism: An Urban Political Ecology of Health and Sustainability

Jeff Rose

Abstract Unsheltered homelessness has increasingly become a standard expectation as part of the contemporary urban landscape. This phenomenon is problematic for multiple reasons. Primarily, unsheltered homelessness is a concern because it represents a very real form of human suffering for the individuals facing homelessness. For the vast majority of people, they very desperately want a more stable and permanent housing option, and their inability to secure stable housing is often a difficult and desperate form of distress. Secondly, for urban community members not facing homelessness, unsheltered homelessness is often perceived as unsightly, and offends their sense of what a city—and a society—should be. Not only is unsheltered homelessness a visible sign of poverty and inequality, but it also disrupts normative visions of public space and public ecologies—seeing unsheltered homelessness represents, in part, a collective failure to produce a public domain that is equitable, inclusive, healthy, and sustainable. Common imagined landscapes of functional urban environment simply do not account for unsheltered homelessness. The ways in which we position unsheltered homelessness, and how we might do so in a more just manner, is the subject of this chapter.

Keywords Homelessness · Urban metabolism · Political ecology

Introduction

People residing in the public spaces of the city—parks, sidewalks, alleyways, waterways, "frags," and elsewhere—appropriate spaces in the public realm (temporarily or otherwise) for survival. In this appropriation of public space, those facing unsheltered homelessness lay bare the socioenvironmental relations that characterize aspects of all human life, including eating, breathing, disposing of wastes, and others. These human metabolic processes for people facing unsheltered homelessness uncovers a justice-focused analysis in which a "right to metabolism" is a fundamental necessity for city life. Here, an urban political ecology approach is developed to consider the

J. Rose (✉)
University of Utah, Salt Lake City, USA
e-mail: jeff.rose@utah.edu

A. Melis et al. (eds.), *Temporary Appropriation in Cities*,
https://doi.org/10.1007/978-3-030-32120-8_9

ways in which a right to metabolism demonstrates that the temporary appropriation of space by those facing homelessness is a biopolitical claim to survival. Before diving into the particular material conditions that define contemporary unsheltered homelessness, it is helpful to first understand how the concept of metabolism has been developed in critical political ecology thought and literature.

Metabolism

Contemporary notions of metabolism most commonly focus on it as a series of material, biochemical interactions that are necessary within a living organism to support and maintain life. Usually, metabolism focuses on transitioning inputs (i.e., food and air) into both energy and cellular matter, in addition to processes associated with the elimination of wastes. While these contemporary approaches remain relevant, it is first helpful to broaden the discussion of metabolic interactions to a more conceptual realm.

Marx and Metabolism

Political, economic, and social philosopher Karl Marx famously considered the relations between labor, capital, and social systems in his extensive 19th Century analyses. Marx understood that not only did capitalism have a tendency to undermine its own economic preconditions for its perpetuation (first contradiction of capitalism), but that capitalism also tends to undermine the environmental conditions for its perpetuation (second contradiction of capitalism). These contradictions, both economic and environmental, demonstrated for Marx the ultimate futility of capitalism, particularly with its concomitant inequalities and human suffering.

Human social life takes place among social institutions, cultural practices, and political economic institutions, but these complex relationships are also entangled with a variety of material cycles, flows of energy, environmental conditions, and resource contexts. This socioecological system, as complex as it is, is also dynamic, changing and unfolding in interrelated ways that are both predictable and impossible to foresee. In the *Grudrisse*, Marx (1973[1858]) understood that these systems—both social and material—metabolize together (White, Rudy & Gareau, 2015). In other words, their transitions and flows of inputs, energies, and outputs dynamically entangle themselves with one another. Marx's historical materialism encouraged a view of reality that is simultaneously historically situated and dynamically unfolding, and necessitated the metabolic interaction of humans and nature:

> It is not the unity of living and active humanity with the natural, inorganic conditions of their metabolic exchange with nature, and hence their appropriation of nature, which requires

> explanation, … but rather the separation between these inorganic conditions of human existence and this active existence, a separation which is completely posited only in the relations of wage labor and capital. (Marx, 1973[1858], p. 489)

Elsewhere, Marx (1978[1844]) outlined how humans constantly do work, or labor, on nature, and in doing so they transform both nature and themselves (and their social institutions). This socionatural metabolism, including labor and nonhuman nature, was for Marx the foundation of history (Swyngedouw, 2009). These ongoing, dialectic metabolic relations between labor and nature are the institutions, relations, and ideologies that that constitute the capitalist mode of production (White, Rudy & Gareau, 2015). Subsequently, scholars have diagnosed a "metabolic rift" in Marx's work, and particularly in *Capital, Vol. I*, where Marx saw capitalism "not only robbing the worker, but [also] robbing the soil" (Marx & Capital, 1977 [1867]), and also in *Capital, Vol. III*, where Marx analyzes the "irreparable rift in the interdependent process of social metabolism" (Marx, 1981[1894], p. 949). In sum, Marx understood socionatural processes as inevitably intertwined but also constantly trying to be separated—materially, discursively, and conceptually—by capitalist ideologies and practices. Metabolism, for Marx, was not the clinical scientific notion that we commonly have today, but an interconnected system including humans, nature, labor, production, ideology, and others, and could be applied to scales much greater than the human body.

Metabolism in Urban Political Ecology

The 20th Century saw the development of considering the city as a living, breathing organism, complete with its own metabolic functioning. Attempts were made to understand the flows of inputs, energies, and outputs at urban and regional levels. While systems-based analyses like Wolman's (1965) "The metabolism of cities," understood various material streams that maintained the life of cities and their residents, these contributions to urban metabolism were based in the scientism of the time. Such analyses did not account for power-laden social systems and their contributions to the interconnected flows of resources, energy, and capital. Political ecology came to critically address these deficiencies.

Political ecology is a body of knowledge and a community of practice that explicitly combines critical political economy with ecology (Robbins, 2011). The social and material production of landscapes and environments – including environmental problems—remains "flush with power relations embedded in the processes of determining what counts as nature, and who counts in society" (White, Rudy & Gareau, 2015: p. 150). While relational, hybridized, poststructural, posthumanist, and other strands have developed to expand political ecology's intellectual and practical depth, the critical political ecology that focuses on Marxist analyses continues to inspire research, engagement, scholarship, and action concerning metabolisms.

The 1990s saw the expansion of political ecology from its roots in rural, agrarian, and natural resource-based settings primarily in the Global South, or the "developing world," to explanation and analyses of definitively urban settings, processes, and engagements. An explicitly urban political ecology developed (c.f., Gandy, 2003; Swyngedouw, 1996, 1997; Swyngedouw & Heynen, 2003), providing critical insight on the ways in which "urbanization is a process which wields nature, society, technology, culture and the city together materially and semiotically through a series of network infrastructures and diverse social power relations moving from the local to the global" (Swyngedouw, 1996: p. x). The city is rife with power, operating across scales in social and material ways. Urban political ecology came to know cities as the sociospatial terrains in which:

> interwoven knots of social process, material metabolism, and spatial form that go into the formation of contemporary urban socionatural landscapes… It is on the terrain of the urban that accelerating metabolic transformation of nature becomes the most visible, both in its physical form and its socioecological consequences. (Swyngedouw & Heynen, 2003: pp. 906–907)

Urban political ecology, both implicitly and explicitly, advocates for democratically governed regulatory structures that combat social, environmental, and socioenvironmental injustices, and caries forward many of Marx's notions of metabolic dynamism that characterizes nature-society relations in the urban realm.

From the Right to the City to the Right to Metabolism

Many urban political ecologists have leveraged the "right to the city" work of critical urban theorist Henri Lefebvre (1996[1968]), a movement that calls for a democratically re-envisioned urban setting. Lefebvre referred to a reappropriation of the decision-making processes and the city's production facilities by its inhabitants. The affirmation and participation of citizens in the future of the city was thought to reduce inequities between citizens, should limit segregation, and foster the emergence of a more inclusive and democratic city. For Lefebvre et al. (1996), the right to the city was "formulated as a transformed and renewed the right to urban life" (p. 158).

Building upon Lefebvre, urban political ecologists tied the right to the city to a "right to metabolism," an entanglement of theory that continues apace today. In their "manifesto for urban political ecology," Heynen, Kaika, and Swyngedouw state:

> Social power relations (whether material or discursive, economic, political, and/or cultural) through which metabolic circulatory processes take place are particularly important. It is these power geometries, the human and non-human actors, and the socio-natural networks carrying them that ultimately decide who will have access to or control over, and who will be excluded from access to or control over, resources or other components of the environment and who or what will be positively or negatively enrolled in such metabolic imbroglios. These power geometries, in turn, shape the particular social and political configurations and the environments in which we live. Henri Lefebvre's "Right to the City" also invariably implies a "Right to Metabolism." (Heynen, Kaika & Swyngedouw, 2006, p.12).

At its core, "the right to metabolism is implicit in the right to the city since the city is the outcome of multiple, continuous, intertwined social and environmental processes and change" (Shillington, 2013: p. 106). What this right to metabolism actually means, and what it implies for the future of urban environments, remains a developing question today.

The power underlying metabolic processes both within cities and across (global) networks of cities and surrounding areas is necessarily a part of this right to metabolism. In essence, "the right to urban metabolism is about the ability of individuals and groups to produce socioenvironmental conditions which create socially and ecologically just living conditions for themselves while at the same time not violating others' ability to do the same" (Shillington, 2013: p. 106).

The construct of metabolism, while primarily deployed in political ecology literature and theory as a metaphor, is also more than just a metaphor. The Marxist metaphor of metabolism describes how urban natures are organized, perceived, and enacted in relation to dominant socioeconomic functioning of cities (Heynen, Kaika & Swyngedouw, 2006). However, functionalist approaches to urban metabolism gain strength with hybridized and relational approaches (e.g., Gandy, 2003; Lawhon, Ernstson & Silver, 2014; Sultana, 2011). Particularly as urban political ecologists embrace embodiment, corporeality, and materiality more broadly, metabolic functioning becomes another contested site of justice (Doshi, 2017). As the spatial and analytical scale of urban metabolic processes moves from the city (with its flows, energies, inputs, outputs, and so forth) to the individual, the embodied sense of metabolism becomes much more materially embodied. The embodiment, of course, is not an abstract mechanism of regulatory individual functioning, but a lived, functional, corporeal, and often emotional series of processes that fundamentally define and prescribe the conditions of life.

Unsheltered Homelessness and the Right to Metabolism

In understanding and defining the parameters around "the right to the city," Mitchell (2003) critically points to the notion that homelessness is perhaps the ideal condition and the ideal geographic setting to consider critically whose rights and what spaces are at stake in the production of urban spaces. "Homelessness has become so prototypically the bellwether of urban justice. If anyone needs the right to the city, surely it is the homeless" (Mitchell, 2003: p. 9). Those facing homelessness are among the most marginalized and stigmatized members of society, and also among the most vulnerable to crime, illness (physical and mental), and overall victimization.

Homelessness, as a condition, includes those large swaths of people who struggle to find consistent housing, but people facing unsheltered homelessness are a subset of this larger demographic, those who, for a variety of reasons, cannot or do not have access to a permanent structure for living. Far from a homogenous population, those facing homelessness are diverse, with the unsheltered population being more likely to have had psychiatric treatment, consume more quantity of alcohol more regularly, and more likely to work in day labor settings (Larsen, Poortinga & Hurdle, 2004).

Unsheltered homelessness is a felt, lived, experienced, ongoing interaction between people and place. It is an intimate, ongoing, embodied interaction between people and nature, or a metabolic relationship as Marx might have termed it. Unsheltered homelessness can be characterized as, in part, an ecological relationship, and subsequently an ecological concern (Rose, 2017). Additionally, a fundamental understanding of urban political ecology is that multiple, interconnected relations of difference and power shape urban ecologies (Doshi, 2017). For instance, homelessness experienced in extreme environmental conditions exemplify the vulnerability associated with homelessness, as demonstrated perhaps most viscerally in severe weather events such as hurricanes, floods, storms, and heat and cold waves (e.g.,Turkewitz, 2017). Even annual seasonal variations (exposure to cold, heat, precipitation, wind etc.) illustrate the ecological interactions and interdependencies taking place for folks without consistent shelter (Pepper & Jocoy, 2013). Perhaps counterintuitively, rates of cold-related mortality actually are higher in regions with fairly mild climates, such as the southern United States and southern Europe, than it is in consistently colder regions (Analitis et al., 2008; Curriero et al., 2002), primarily due to issues of short-term daily *temperature range*, as opposed to *absolute temperature*. Hypothermia likelihood is highest not necessarily when temperature is the coldest, but when nighttime low temperatures are in marked contrast with relatively warm daytime temperatures (National Coalition for the Homeless, 2010). A consequence of these phenomena is that spring and fall seasons are often the most problematic for individuals without shelter, when temperature and precipitation are often least consistent in the United States (Pepper & Jocoy, 2013). As climate change increases global temperatures, exposure to extreme heat and the necessity of access to water is an increasing environmental justice concern(DeMyers, Warpinski, & Wutich, 2017), with increasing reports of increasingly difficult summer conditions for those living outdoors (e.g. Gee, 2017, July 24th). People living without consistent shelter are disproportionately vulnerable to extreme heat and heat-related deaths than people living elsewhere (Harlan et al., 2012). Many of the conditions often associated with unsheltered homelessness—mental illness, drug addiction, alcoholism, etc.—make these people further prone to making poor personal health decisions during difficult environmental conditions, exacerbating the concerns.

These literatures support the notion that local environmental conditions (in this case, weather and climate) directly impact experiences of homelessness. "Of all the hardships of homelessness, it is the fundamental physical circumstances—exposure to the elements, the struggle to keep clean, the discomfort of bedding down on concrete—that are often the most piercing" (Gee, 2017, July 24th). At first blush, this conclusion is not surprising, as local environmental conditions affect everybody's lives, to varying degrees. However, for folks facing homelessness, these daily environmental conditions are of vital importance, whereas many urban community members can retreat to residences, work spaces, vehicles, or other (private) spaces to maintain health and wellness, as well as physical comfort and conditions for success. These are spaces where we can limit our exposure to (challenging) environmental conditions, where we can mediate our nature-society relations as we see fit. Individuals facing homelessness are much more conscribed in their options to make similar choices.

While urban political ecologists often think of metabolism occurring at the scale of the city itself, it is clear from this presentation of unsheltered homelessness that metabolism also occurs at the scale of the body as well (e.g., Shillington, 2013). Intimate interactions and interplays between people and nature occur with such immediacy and such intensity for those facing unsheltered homelessness that their ongoing metabolic functioning necessitates both space and particular socioenvironmental conditions within the urban milieu. Gandy (2004) identifies metabolism as the "circulatory processes that underpin the transformation of nature into essential commodities such as food, energy, and potable water" (p. 374). While every living creature takes part in these processes to varying degrees, those facing unsheltered homelessness do so in such an obvious, visceral, and publicly visible manner.

The ways in which we *produce* urban environments, in this light of unsheltered homelessness, moves from common perspectives of aesthetics and efficiencies to considerations of elemental functionality and survival. The ways in which we envision, develop, regulate, legislate, police, and otherwise control our urban spaces has immediate ramifications for those living outdoors, on sidewalks, street corners, parks, and waterways. Their right to metabolism is dependent upon their ability to regulate the flows of inputs, energies, and outputs, and urban environments' functionality and livability become fundamental questions of survival. For folks facing unsheltered homelessness, our urban socionatures are their homes and livelihoods. Or, as Shillington (2013) concludes, "individual metabolism is also linked to the production of urban nature" (p. 110).

Homelessness, Temporary Appropriation, and Urban Metabolism

Contemporary unsheltered homelessness is nothing if not about the temporary appropriation of urban public space (Marcia Johnson, 2008; Mitchell, 2003; Mitchell & Staeheli, 2006; Rose, 2017; Schein, 2012). The absence of consistent access to private space, and therefore being forced to live one's life in public space, is a defining feature of homelessness. Urban homelessness becomes about the ephemeral mobility in and out of spaces in the city. Public space, as opposed to private space, being governed in common, means that folks facing homelessness (and others) can only participate and exist in public space temporarily. For instance, parks, perhaps the most open and inclusive of urban spaces, tend to refer to people interacting in these spaces as "visitors," temporary users and participants. Ideally, public space is inclusively available to everyone at all times, but practically, individuals facing homelessness are regularly told to "move along" (e.g., Coleman, 2017; Wolch, Rahimian & Koegel, 1993) as they live their daily existences in public, despite the nascent evidence that many non-homeless park users are perhaps more tolerant of unsheltered homelessness than might be expected (Rose, 2019). Public space being inclusive and accessible is, of course, contingent and conditional. Often the contingency of public space's accessibility is dependent upon one's class (among other intersecting aspects of identity), and individuals facing homelessness are simply vulnerable enough that a variety of political, economic, social, and legal forces can and do displace them with regularity (Mitchell, 2003).

Unsheltered homelessness differs only in that it provides a higher degree of publicness, of vulnerability, and of being exposed and visible for the world to see. In not residing in shelters, those facing unsheltered homelessness simply have no (private) space of retreat, and no recourse but to live the entirety of their existence in public space, including eating, sleeping, socializing, resting, and disposing of wastes. And while reports of extensive and longer-term "tent cities" across a wide variety of urban environments are on the rise (Hunter et al., 2014; Sparks, 2017), the vulnerability, precarity, and (relatively) short-term intermittent nature of these encampments is unquestioned. Spaces in the city, from the sidewalk to the street corner to the town square to the municipal park, are only available to those facing homelessness in fleeting instances, and at any time these spaces can be enclosed, making them instantaneously unavailable.

These temporary appropriations of public space by those facing unsheltered homelessness are vital in the production of a right to metabolism in the city. Metabolic functioning necessarily has spatial components, and at the most basic analysis, there must be the physical space for the flows of inputs, energies, and outputs. When folks facing homelessness make claims to exist in the urban fabric, they are making claims to metabolize, or as Shillington (2013) notes, "appeals for clean water, sanitation, food, etc. are not just claims for the right to the city, but also for the rights to urban metabolism" (p. 106). And while a right to metabolism surely is a valid claim for all community members, it is those at the margins of society, those who are most vulnerable, whose claim must be amplified most clearly. "Movements of marginalized and precarious urban dwellers who resist pressures… and, instead, claim the right to inhabit the city, may be conceptualized as protagonists of a struggle for the right to metabolism" (Olsen, Orefice & Pietrangeli, 2018: p. x).

The need to claim and reclaim a right to metabolism in contemporary urban environments is a clear extension of Lefebvre's right to the city, but also a necessary corollary of Marx's contradictions of capitalism. As metabolic functioning is vital at both the urban and bodily scales, movements to reclaim a right to metabolism underscore the analyses that capitalism is actively undermining its own conditions for existence, whether that is understood at the level of the laborer or the nonhuman material environment. Those facing unsheltered homelessness quite clearly embody the difficulties—simultaneously in terms of personal health, community health, and environmental sustainability (Rose & Johnson, 2017)—brought forward from the contradictions of capitalism. But these individuals and communities are not the only members of society suffering the consequences; they are unfortunately only the most vulnerable leading edge of a political economic system whose contradictions are likely to increase suffering on much larger segments of the population. Advocating for a right to metabolism is a vital path forward for making urban space, and urban nature-society relations more accommodating to people and more ecologically functional:

> The (re)production of (everyday) urban natures and the ways in which they challenge and alter other urban 'natures' is not only integral to the right to the city demands – and therefore, the right to metabolism – but also critical in the creation of the city as an oeuvre which reflects the struggles of ordinary citizens in their everyday lives. (Shillington, 2013: p. 110)

References

Analitis, A., Katsouyanni, K., Biggeri, A., Baccini, M., et al. (2008). Effects of cold weather on mortality: Results from 15 European cities within the PHEWE project. *American Journal of Epidemiology, 168*(12), 1397–1408.

Coleman, J. R. (2017). Diary of a homeless man. In: *Housing the homeless* (pp. 37–52). Routledge.

Curriero, F. C., Heiner, K. S., Samet, J. M., Zeger, S. L., et al. (2002). Temperature and mortality in 11 cities of the eastern United States. *American Journal of Epidemiology, 155*(1), 80–87.

DeMyers, C., Warpinski, C., & Wutich, A. (2017). Urban water insecurity: A case study of homelessness in Phoenix, Arizona. *Environmental Justice, 10*(3), 72–80.

Doshi, S. (2017). Embodied urban political ecology: Five propositions. *Area, 49*(1), 125–128.

Engels, F., & Tucker, R. C. (1978). *The Marx-Engels Reader*. In: R. C. Tucker (Ed.). New York.

Gandy, M. (2003). *Concrete and clay: Reworking nature in New York City*. MIT Press.

Gandy, M. (2004). Rethinking urban metabolism: Water, space and the modern city, *8*(3), 363–379 (Online). Available from: https://doi.org/10.1080/1360481042000313509.

Gee, A. (2017). Chronicling homelessness: The summer heat takes a brutal toll |US news|. *The Guardian*. 17 July (Online). Available from: www.theguardian.com/us-news/2017/jul/24/homelessness-summer-us-southwest-phoenix-los-angeles.

Harlan, S. L., Declet-Barreto, J. H., Stefanov, W. L., & Petitti, D. B. (2012). Neighborhood effects on heat deaths: Social and environmental predictors of vulnerability in Maricopa County, Arizona. *Environmental Health Perspectives, 121*(2), 197–204.

Heynen, N., Kaika, M., & Swyngedouw, E. (2006). In the nature of cities: Urban political ecology and the politics of urban metabolism (Online). London, UK: Routledge. Available from: https://urbanforensics.files.wordpress.com/2012/09/inthenatureofcities.pdf.

Hunter, J., Linden-Retek, P., Shebaya, S., & Halpert, S. (2014). *Welcome home: The rise of tent cities in the United States*. National Law Center on Homelessness & Poverty, Allard K. Lowenstein International Human Rights Clinic, Yale Law School. https://www.nlchp.org/documents/WelcomeHome_TentCities. Last Accessed June 24, 2016.

Larsen, L., Poortinga, E., & Hurdle, D. E. (2004). Sleeping rough: Exploring the differences between shelter-using and non-shelter-using homeless individuals. *Environment and Behavior, 36*(4), 578–591.

Lawhon, M., Ernstson, H., & Silver, J. (2014). Provincializing urban political ecology: Towards a situated UPE through African urbanism. *Antipode, 46*(2), 497–516.

Lefebvre, H. (1996). *Writings on cities*. Malden, MA: Blackwell.

Marcia Johnson, E. (2008). An investigation into pedagogical challenges facing international tertiary-level students in New Zealand. *Higher Education Research & Development, 27*(3), 231–243 (Online). Available from: https://doi.org/10.1080/07294360802183796.

Marx, K. (1973). *Grundrisse*. In: M. Nicolaus (Ed.). New York, USA: Vintage.

Marx, K. (1977). *Capital Volume I*. In B. Fowkes (Ed.). New York, USA: Vintage.

Mitchell, D. (2003). The end of public space? *The Right to the City—Social Justice and the Fight for Public Space*, *5608*, 118–160 (May 2014) (Online). Available from: https://doi.org/10.1111/j.1467-8306.1995.tb01797.xa.

Mitchell, D., & Staeheli, L. A. (2006). Clean and safe? Property redevelopment, public space, and homelessness in downtown San Diego. *The Politics of Public Space, 2006,* 143–175.

National Coalition for the Homeless. (2010). *Winter homeless services: Bringing our neighbors in from the cold* (Online). Available from: http://www.nationalhomeless.org/publications/winter_weather/Winter_weather_report.pdf.

Olsen, E. S., Orefice, M., & Pietrangeli, G. (2018). From the 'right to the city'to the 'right to metabolism'. In *Housing for degrowth* (pp. 33–43). Routledge.

Pepper, D. A., & Jocoy, C. L. (2013). A climatological analysis of emergency homeless shelter openings in Long Beach, California, USA. *Applied Geography, 37,* 168–175.

Robbins, P. (2011). *Political ecology: A critical introduction.* Hoboken: Wiley.

Rose, J. (2017). Cleansing public nature: Landscapes of homelessness, health, and displacement. *Journal of Political Ecology, 24*(1), 11–23.

Rose, J. (2019). Unsheltered homelessness in urban parks: Perspectives on environment, health, and justice in Salt Lake City, Utah. *Environmental Justice, 12*(1), 12–16.

Rose, J., & Johnson, C. (2017). Homelessness, nature, and health: Toward a feminist political ecology of masculinities. *Gender, Place & Culture, 24*(7), 991–1010.

Schein, R. (2012). Whose occupation? Homelessness and the politics of park encampments. *Social Movement Studies, 11*(3–4), 335–341.

Shillington, L. J. (2013). Right to food, right to the city: Household urban agriculture, and socionatural metabolism in Managua, Nicaragua. *Geoforum, 44,* 103–111.

Sparks, T. (2017). Neutralizing homelessness, 2015: Tent cities and ten year plans. *Urban Geography, 38*(3), 348–356.

Sultana, F. (2011). Suffering for water, suffering from water: emotional geographies of resource access, control and conflict. *Geoforum, 42*(2), 163–172.

Swyngedouw, E. (1997). Power, nature, and the city. The conquest of water and the political ecology of urbanization in Guayaquil, Ecuador: 1880–1990. *Environment and planning A, 29*(2), 311–332.

Swyngedouw, E. (2009). The antinomies of the postpolitical city: In search of a democratic politics of environmental production. *International Journal of Urban and Regional Research, 33*(3), 601–620.

Swyngedouw, E. (1996). The city as a hybrid: On nature, society and cyborg urbanization. *Capitalism Nature Socialism, 7*(2), 65–80.

Swyngedouw, E., & Heynen, N. C. (2003). Urban political ecology, justice and the politics of scale. *Antipode, 35*(5), 898–918.

Turkewitz, J. (2017). For Houston's homeless, a terrifying night under siege by Hurricane Harvey. *New York Times* (Online), 27 August. Available from: https://www.nytimes.com/2017/08/27/us/hurricane-harvey-homeless-houston.html.

White, D., Rudy, A., & Gareau, B. (2015). *Environments, natures and social theory: Towards a critical hybridity.* Macmillan International Higher Education.

Wolch, J. R., Rahimian, A., & Koegel, P. (1993). Daily and periodic mobility patterns of the urban homeless. *The Professional Geographer, 45*(2), 159–169.

Wolman, A. (1965). The metabolism of cities. *Scientific American, 213*(3), 178–193 (Online). Available from: http://www.jstor.org/stable/24931120.

Temporary Appropriation in Shanghai and Hong Kong: Two Study Cases Assessing the Resilience of Women Faced with the Lack of Affordable Housing

Alessandra Lai

Abstract The scholastic notion of 'temporary appropriation' of the public space refers to a wide range of urban scenarios that evoke conditions of either spatial, temporal or socio-economic disorder. Such disorder is manifested by the frequent contrast between the planned purpose of a public space and its actual use. Often, the unforeseen habits of the users of a place obscure the planned design narrative, adding a juxtaposed layer to the urban life. Analysis of the way people make use of a public space can reveal the contrasts by which the order designed by the planners becomes lost under a disorder or "not coded order" practiced by the users. Through direct observation via qualitative and quantitative investigation of diverse sources of the cause of 'temporary appropriation', this chapter investigates two phenomena in relation to its occurrence in public spaces in China. The first is the characteristic spreading of various domestic functions (such as cooking and laundry) onto the sidewalks of Shanghai. The second is the widespread use by Filipino maids of the squares and bridges of Hong Kong during their days off. The main aim of this essay is to highlight how such 'user invasions' reflect the resilience of these largely female communities in order to effectively and efficiently face issues related to affordable housing.

Keywords Temporary appropriation · Appropriation · Social sustainability · Resilience · Housing affordability

Introduction

Although the term appropriation has a well defined meaning in the jurisprudential sphere, it still remains fairly loose in the field of urban design and in particular of urban sociology. Not only there is no agreement, here, about its meaning, but neither is there an established basis for the evaluation of the practices attributable to appropriation. We agree with Lara-Hernandez, Caputo and Melis (2017) that it is

A. Lai (✉)
University of Sassari, Sassari, Italy
e-mail: alelai2014@gmail.com

A. Melis et al. (eds.), *Temporary Appropriation in Cities*,
https://doi.org/10.1007/978-3-030-32120-8_10

dubious referring to appropriation as either an informal or an illegal activity in the public space, since that space is intended to for citizens and for their benefit.

Before describing the case studies, it may be useful to introduce the conceptual references that frame the study, its interpretation and its narration.

The core of the work comprises the analysis of two representative city areas, one in Shanghai and one in Hong Kong—taken as study cases for the analysis of Temporary Appropriation (TA) in relation to the resilience of women faced with the lack of affordable housing, and the ways in which such appropriation relies on the informal use of streets and squares.

Given the relative novelty of the subject, this chapter focuses on the interest shown in daily news, on data available via the newspapers, in addition to direct observation. We aim to identify the potentials in the marginalized communities to produce a temporary new urban landscape by re-appropriating or re-configuring the existing spaces. Thus, we question whether the act of appropriating can be identified *as a form of creative usage of space or whether it is merely a survival* act. In this sense we highlight different ways where TA is intended as a consequence of the necessity of adapting human needs of a city that deprives part of its population access to respectable housing conditions.

It will therefore advocate the proposal of including the activities related to housing in the list of the three categories provided by Lara-Hernandez and Melis (2018) where they group the activities taking place in public spaces that are relevant for the phenomena according to their social, cultural and environmental context-specific nature, according with several scholars such as Crossa (2009), de la Torre (2015), Kim (2013), just to name a few. In fact, this study would give its contribution on the attempt to frame the otherwise ambiguous sphere of TA practices, by claiming that there are different ways in which TA reflects the aim to overcome urban restrictions, such those linked to housing affordability: as well as the activities related to the economy, which are extended from the place of working to the public realm (street), they are an example of TA.

Finally, this work explores the possibility to use the activities related to housing affordability, as expression of TA, as a strategic key to enrich the discussion on micro-scale urban public spaces like streets and squares. The study cases here described represent a rich context to approach some urban phenomena that can be of great impact, particularly in terms of social sustainability, on the micro-scale urban design and management.

Temporary Appropriation as a Strategy to Address Housing Affordability Issues in Big Cities

According with several scholars such as Bay and Lehman (2017) and Ekawati (2015), since the urban population is expect to increase to seventy per cent by 2050, the informal use of public spaces argues to be a more sustainable urban model. In particular we agree with Ekawati (2015) when points out that it's not only the risk of scarcity of urban open space in the high-density cities that will increase the multidimensional use of the streets, but mostly the scarcity of housing space. We, as humans are committed with the appropriation and construction of our surroundings (Yory, 2011). That's why within the field of environmental psychology, the term appropriation is described as a temporary phenomenon that implies a dynamic process of interaction between the individual and its surroundings (Korosec-Serfaty, 1976). This interaction or reciprocity exists thanks to the evidence that we do not just adapt our behavior and features to the environment conditions: we have also the ability to think creatively, to imagine novel solutions to survival threats and overcome obstacles that comes from the environment we live in, adapting its features to our needs.

There are social, urban and cultural conditions in which individuals or groups of them are challenged to appropriate instead of just using public spaces in order to manage their everyday life (Lefebvre, 1992). Therefore, appropriation can be seen as an inborn necessity of humans, that can be expressed through activities that occur in the urban landscape and it conforms the urban identity through individual and collective practices (de la Torre, 2015; Pol & Iñiguez, 1996; Pol Urrútia, 2002; Vidal-Moranta & Pol Urrútia, 2005). But while appropriation is a broader term, its temporary variation refers specifically to public spaces, so TA become the temporary act in which people use public spaces to carry out individual or collective activities other than the purpose that the space was originally designed for (Fonseca Rodriguez, 2015).

As Nasution (2015) pointed out, it is mostly marginalised people that are continuously reinventing their understanding of space and territory in the public domain. Resiliency and creativity emerge to commodify every possible inch of available space.

Since the formation of a city is a necessity for proximity to economy and social relation, it is obvious that when it becomes denser with people and activities, space becomes a precious commodity. Thus, a city consciously commodifies and regulates the spaces.

The study cases presented in this work focus on temporary appropriation as a sustainable urban model adopted in high-density cities as Shanghai and Hong Kong to face the scarcity of housing space. They will be described in two independent section, but analyzed in a common one, with a particular focus on their contribution to social sustainability in a urban environment and on the common features that candidate the activities related to housing to integrate the list of the three categories that delineate TA.

Shanghai Creation of an Outdoor Domestic Landscape to Deal with the Limited Availability of Housing Space

Already in 1926, the writer Aldous Huxley, highlighted that none of the cities he had ever visited so overwhelmingly impressed him with its teeming humanity as Shanghai. He had never had such an impression of dense, rank, richly clotted life (Huxley, 1926). Xiaolong (2013), writer and scholar of Chinese literature, noted that in Shanghai the everyday life is enriched by all the activities that take place in the street. "You become parte of the alley and the alley become part of you". That's because people use to appropriate the immediate vicinity of their habitat making a semipublic space of the alleys and sidewalks (See Fig. 1).

Both the admired gaze of Huxley, as well as the nostalgic one of Xiaolong, highlight how the practices of appropriation of public space in the vernacular alleys of Shanghai, have generated images and meanings that are now universally associated with a piece of social and cultural history and Shanghai politics that is disappearing (Cappelletti, 2014) and that finds in the changing configurations of the urban scene the most effective representation of its decline.

As well as Huxley and Xiaolong, other writers and scholars, such as Wang (2008) or Li (2014), describe the urban housing form that has characterized the city for over 150 years based on the repetition of a grid made of houses aligned, one next the other, along alleys (Lilong 里里: the Shanghai term for Longtang 里里).

Fig. 1 The use of sidewalks in Shanghai (Photograph by Poddie, L. 2014)

Originally the alleyways homes were single-family houses, but when the city's population started rising up, in the early twentieth century, as prices grew with the housing demand, many tenants remodeled their houses and sublet rooms to subtenants. The inner courts, formerly private open spaces, were partitioned and wings and tentacles grew all over around them (Li, 2014).

In order to deal with the limited availability of housing space and natural light and air, the residents, mainly low income workers, started to invade streets by bringing out their domestic spaces. It is both the potential and danger. At a first sight, this condition makes the alleys look chaotic and unappealing, but when we move to another level of analysis, we can argue that it makes the alleys lively and animated (or even productive). In this new space, people become aware of each other activities. The appearance of the activities creates a greater possibility for cooperation (Nasution, 2015). The range of activities that take place in the lanes goes from recreational activities such as cooking, mahjong, laundry drying, outdoor dining, to commercial activities such as street vendors, barbers, grinders.

This chapter focuses on the domestic one. The activities that take place in the alleys are connoted by daily frequency. They don't require a radical change of the street design: there is just one element that is permanent: an industrial concrete sink attached to the external wall of the house. The burners/flames are often allocated in a wooden freestanding sideboard, that can be moved easily from inside to outside. The public net of suspended electric cables is used to hang wet clothes. Chairs, benches or lounger are often outside among many other stuffs (see Fig. 2).

All this equipment occupies the alleys in a variable arrangement, offering a never fixed layout/configuration.

It is interesting to note that all the sets of informal arrangement do not prevent "strangers" to use the space, nevertheless they affect the walks through the alleys, enriching, as said at the beginning, the experience, otherwise really "flat" or in some cases even scary, as in the residential narrow streets it uses to be. It's not just going through a neighbourhood: it's going through another people life. A feeling of safety arises thanks to the mutual control.

Temporary Social Landscape in Hong Kong

The second case of study concerns the phenomenon of appropriation of public spaces in the city of Hong Kong, limited to the practices of maids, especially Filipino, who use, since the 1980s, to gather every Sunday, their day off, in several squares of the city. For the purpose of this chapter, the observations focus on the phenomenon that takes place in the covered square below the headquarters of the HSBC (The Hong Kong and Shanghai Banking Corporation), designed by the English firm Foster + Partners. The suspension structure that characterizes formally and structurally the building has also implications at ground level, allowing the entire building to be jacked up on steel legs to free the area beneath for the public plaza. It represented a significant gesture in high-density Hong Kong and it continues to be a well-used

Fig. 2 Appropriation of alleys in Shanghai (Photographs by Lowen, J. 2015)

public space space, taking on a welcome lease of life as a sheltered carpet for picknicking Filipino maids at the weekend, who take advantage of the air-conditioning leaking out of the bulging glass ceiling above.

Foreign domestic helpers started arriving to Hong Kong in 1969, encouraged by local governments, such as the Filipino one: the migration eases unemployment at home and provides a source of income via remittances (Kuo, 2014).

In the 1980s, Filipino domestic workers began congregating every Sunday in the Statue Square, which, perhaps as a reflection of the more democratic ethos of the post-war era, become less ceremonial and more of a gathering place, with sheltered seating areas and fountains filled with abstract artworks (DeWolf, 2019). See Fig. 3.

In 2003, after a long discussion, as part of efforts to protect the local workforce, the Hong Kong government concluded that foreign domestic helpers should not be allowed to live away from their employers' residence "as such arrangements would affect the job opportunities of local domestic helpers". It thus made the provision of a full-time, live-in domestic service a "fundamental principle of the foreign domestic helper policy" and decided employers should only be able to hire foreign domestic helpers if they were able to house them in their homes (Siu, 2018). As a consequence of this rule, the Filipino maids are expected to work all day long. Their duties include cooking, household chores, babysitting, child care and elderly care. Given that most Hong Kong homes are spatially-challenged, many of them sleep in tight quarters with a very few chances of having a proper rest. The desire to live away

Fig. 3 Sunday's urban scenarios in Hong Kong [photograph by author (left), Lara-Hernandez, J.A. (right)]

from employers is therefore strong and growing, even with Hong Kong's housing crunch (Miquiabas, 2015).

The foreign presence of the domestic helpers in Hong Kong can only be perceived through the use of the public spaces of the city (squares, parks, roads) as meeting places, exchange of information, economic activities, because they spend the rest of the time in the houses where they work. From the point of view of distribution logics and settlement patterns, it should be noted that, if the dwelling is in the workplace, as in the case of domestic work, the visibility of migrants in the residential areas is lower, while the relative visibility is greater to some public spaces where they meet with their fellow citizens.

Although they are not designed to be a form of protest or a way to attract attention, the size of these spontaneous gatherings, given the numbers (there are around 350,000 foreign maids working in Hong Kong according to Miquiabas) has attracted the attention of the media. Their weekly recurrence and localisation invariance does not leave indifferent.

Every Sunday morning a new layer is added to the public space and the formally designed one is removed. Due to its characteristics the new one constitutes a new urban landscape with very strong social connotations. The financial and commercial district of Hong Kong changes radically. Women sit, sing, watch television, prepare packages to send to families, eat and sleep. The Spanish artist Marisa González, in the project "Female (Open) Space Invaders" (2010), has documented the life of Filipino maids in Hong Kong. Her photographs show how these women reconstruct urban spaces to form unique social communities in Hong Kong. Nothing is permanent. They reshape the square under the suspension structure of the HSCB adding cardboard boxes that give a structure to the empty space of the square with temporary open-air rooms, where the exhausted workers stay seated most of the time, to recover the energies lost during the week. Once the day gets to the end, they move away everything, leaving the space ready to host the frenetic fluxes of the Financial District on the next morning.

As well as it happens in Shanghai, it is a form of resilience rather than resistance. Foreign maids do not come together to show discomfort or talk about a problem, which undoubtedly exists, but to meet a common need: to find a space in which to shelter and enjoy themselves on the day of rest, for all of Sunday, in which they are forced to leave the apartment where they serve and live.

Sunday is the day in which each person transports their habits from their home towns to Hong Kong (Gonzalez, 2012).

In the absence of opportunities, the community of domestic workers can rely on their social capital to face the disadvantages of marginalization. Social capital is represented by the resources that are accessible through social contacts, social networks, reciprocity, norms, and trust (Bourdieu, 1986; Coleman, 1988; Field, 2003; Klienhans, Priemus, & Engbersen, 2007; Putnam, 2000) which is achieved by being a member of certain social network (Tunas, 2008). According to Tunas (2008) frequent interaction and long-term relationship are more likely producing a more intense social capital, since they foster strong social cohesion, sense of togetherness, social responsibility and mutual control (see Fig. 3). Through a strong social capital, the community tends to be more resilient and able to take advantage of the available resources. In fact, social capital is one of the best assets of the marginalized community (Tunas, 2008).

As well as the Shanghainese inhabitants of the Lilong, due to limited availability of spaces in Hong Kong, the Filipino maids have learnt to produce temporary informal spaces over formal spaces, by the addition of a layer that is adaptable, changeable and ephemeral. A movement space or recreational space is turned into a productive space. Such condition blurs the inside and outside. And paradoxically, as they consume formal spaces, they also produce informal spaces. Time and event are the only things that separate it. The resiliency of the domestic workers can turn the condition of lacking resources into abundance of resources. The space itself is not embedded by any specific symbolical value. However, the struggle of these workers brings specific common values to the space (Nasution, 2015).

Temporary Appropriation and Domestic Landscapes

The social aspect of sustainability, has been the last one to be taken in consideration, long after the economical and environmental ones, and only has been seriously considered after the year 2000 as an integral part of sustainability that should distinctively debated (Mehan & Soflaei, 2017). As a result, there are still uncertainties in definition, criteria and measurement system until now and therefore there is a little consensus over the definition of social sustainability. For the purpose of this work, it seemed more coherent to approach the concept from an urban design perspective, which usually refers to the physical aspects while evaluating the impacts of built environment design on social sustainability (Secher, 2014). Littig and Griessler (2005) suggest that social sustainability means the satisfaction of basic human needs and the subsequent continuation for future generations. Where the basic human needs are intended as those in the Maslow hierarchy. Mehan and Soflaei (2017) listed the urban social sustainability indicators as following: future focus (means that social sustainability is primarily about valuing and protecting positive aspects of cultures (McKenzie, 2004) also for a generation to come; satisfaction of needs; socially cohesive and physically integrated urban unit (means that physical setting is considered as being conductive to social cohabitation, socially integrated and promoting well being and quality of life of the diverse group).

The spontaneous and informal appropriation, and consequently temporary change, of the formal built environment, carried out by the Shanghainese inhabitants of the lilong or the Filipino maids in Hong Kong, gives a significant contribution to the debate on urban social sustainability. It shows how intangible concepts such as identity, sense of place and community stability and security are not just goals to achieve thanks to a "properly" designed environment, but also resources that can reproduce themselves—future focus—thanks to actions taken on the built environment.

Both Shanghai and Hong Kong face all the issues related to the megacities. This work focuses in particular on those related to the poverty of affordable housing and the solutions implemented by women, both in groups or individually, to gain extra space to carry out some domestic activities and therefore satisfy physiological needs, psychological needs and safety needs.

In the case of Shanghai, this aim results in the occupation of the alleys next to the houses. In the case of Hong Kong, it results in the weekly occupation of some public spaces, mainly squares, that offer some basic conditions of comfort (such as shelter that provide shadow or protect from rain).

Such appropriation of formal spaces to informal spaces requires a strong degree of social cohesion. Both cases show that, being marginalized by the formal city, people are forced to utilize available resources, both physical space, and mental space. However, in such process there is constant negotiation between the formal system and the informal activities. Although, in the cases of Hong Kong, the contribution of the formal system is more apparent than in the Shanghai ones, where on the other side, temporary appropriation strategies show the strength of the negotiation between privately owned spaces and public spaces (Nasution, 2015).

Still, as Nasution (2015) pointed out, marginal space intelligence narrates how space can be collectively produced, embedded with meaning, as well as, it motivates its inhabitants to move forward with certain goals. Spaces are a place for social struggle as well as celebration of the collective goals. They are places where the lack of resources is overcome by the formation of collective goals, dialog and negotiation, and the complex multi-layer of space. We can agree with him when states that the absence of resources becomes the main resource.

The frequency that characterizes TA is also a significant factor. Indeed, according with Kyle, Jun, and Absher (2014) activities, which occur in the urban space, constantly repeated over certain periods of time, help to construct the identity and bonding between place and individuals. In addition, researchers of environmental psychology, such as Vidal-Moranta and Pol-Urrútia (2005) argue that the place that is appropriated is considered as a factor of continuity and stability of the self (it is true in what was observed in Shanghai) in parallel to another factor that contributes to the identity and group cohesion, as the Filipino maids behavior proves.

According with Lara-Hernandez and Melis (2018) conclusion, TA takes place in urban public spaces in which an individual or group of individuals feels culturally identified. However, the case of study of the domestic workers in Hong Kong suggests that the appropriation occurs, first, in spaces that inspire, thanks to some physical features (or just proximity as shown by Shanghai alleys), other uses than the one they were intended for, by designers. Therefore, the social cohesion, based on the group homogeneous identity and shared problems, gives the strength to implement the informal space over the formal one. Finally, the repetitiveness of the practice in the same specific place activates the sense of belonging, whose nature is predominantly cultural, so that people feel culturally identified in that place.

The urban appropriation strategies just mentioned can be considered as socio-cultural narratives. They are free from symbolism or other contents or aspiration, so, rarely pure. They don't want to set and example or send a message. Women just operate choices for their better living.

The domestic workers have developed a common understanding to resist their marginalized conditions, either the alienation by the city or the lack of privacy in their employers' house, and work collectively to achieve a common goal (Nasution, 2015). The appropriation of public space contribute to overcome human alienation (Graumann, 1976).

Dempsey, Bramley, Power, and Brown (2011) argued that social sustainability, within the urban context, strongly relies on two main aspects: social equity and sustainability of community.

Social equity is frequently associated and measured by the accessibility to key facilities that satisfy basic needs within a determinate area (Burton, 2000; Barton, 2002). Nevertheless, social equity is also commonly associated with the inclusion of all members of society without distinction. The citizen's appropriation of the public realm is widely perceived, from the scholars' point of view, as an opportunity to overcome urban segregation and social exclusion (Pol Urrútia, 2002).

Sustainability of community is related to the formation of a community and measured by the level of social interactions between its members. When people, living in a determinate area, do not socially interact among them, they cannot be identify as a community, they are just individuals living separate lives with little sense of pride or place attachment (Dempsey, 2006).

Studies carried out by Seaman and McLaughlin (2014) suggest that the more interaction between individuals and their outdoor environment, the higher levels of place attachment and sense of community. While the Filipino maids use of gathering every Sunday gives an emblematic representation of these arguments, the projection of some domestic activities, such as cooking or laundry, in the alleys of Shanghai seems to give a opaquer prove of social interaction or cohesion. However, the coexistence, along such narrow streets, of intimate, domestic and familiar activities, would not exist without an implicit negotiation, collaboration and sense of spatial adaptability and community.

Both cases can be also recognised as expression of spatial intelligence since they are characterized by negotiation, spatial adaptability, collaboration and collectivity (Nasution, 2015). The Shanghainese—but it's a common phenomenon that interest many other cities - practice of sharing with other people common spaces to carry out domestic activities, doesn't seem very different from the co-housing principles, as they're designed for European new residential buildings. Indeed, the use of the lanes by the Lilong inhabitants of Shanghai to perform domestic functions, expelled from the limited spaces of the houses, has many elements in common with the co-housing and co-working experiments, that in the last ten years are spreading more and more all over the world, or with the structure of mountain village typical of the most remote Chinese housing culture. They are all based on sharing spaces or equipment to divide the costs in terms of money, to free some spaces in houses that are getting smaller, to rely on a stable community to feel safe and avoid human alienation.

Conclusions

In order to support the inclusion of the activities related to housing in the list of the three categories provided by Lara-Hernandez and Melis (2018) to frame the sphere of TA practices in public spaces - activities related to leisure; activities related to economy; activities related to religion - it might be useful to call back some concepts that have been highlighted during the illustration of the two study cases, among other informations. The adherence of the activities, widely described for each of the two cases, to the definition of temporary appropriation will confirm that as well as the activities related to the economy, which are extended from the place of working to the public realm (street), the activities related to housing, which are extended from home to the public realm (street or square), are examples of TA. In fact, those activities:

- Represent a temporary act in which people use public spaces to carry out individual or collective activities other than the purpose for which spaces were originally designed for (Fonseca Rodriguez, 2015);
- Hint the imaginary, symbols and rituals that people's value have towards a specific place, building or urban element and what they mean to them (Rapoport, 2005);
- Express an interaction between the individuals (or collectively) and their city through specific activities happening in the urban landscape (Lara-Hernandez, Melis, & Caputo, 2017);
- Affirm the citizens' right to fully use and manage their everyday life within the urban environment (Lefebvre, 1992);
- Happen in spaces where people feel identify (Vidal-Moranta & Pol Urrútia, 2005; Martínez, 2014);
- Are characterised by the informality of the action;

- Emphasizes that TA relies on both the configuration of the built environment and the meanings and implicated relationships (people's values, behaviours and actions) of the social life in the city (Martínez, 2014).

According to the table elaborated by Lara-Hernandez, Melis, and Caputo (2017), a general description of the activities related to housing is suggested, as well as a set of sub categories.

Activities related to housing include any activity in which a person or group use the public space in order to overcome the lack of affordable and decorous domestic space. They can be split in two sub categories: housework and rest. Each of them includes a set of activities that can be carried out by individual or collectively. Housework means cooking, laundry, dish-washing, hair-washing. Rest is referred to eating, resting, reading and watching videos on tablet or smartphone.

This study has consciously jumped over the implications of the drivers of urban transformation in Shanghai, where the neighbourhood life just described is going to disappear within the next few years, if not earlier. It would deserve a proper space.

At the same time, for what concern Hong Kong, activists and media are challenging the government in order to change the regulation that force domestic workers to live and work in the same place, where they seem to be considered almost as slaves. As well as in the case of Shanghai, this study doesn't go deeper into the topic, because it would mean telling another story.

It should be noted, however, that in both cases the measures in progress to do not condemn TA practices, rather they work on eliminating the causes that generated them.

References

Barton, A. C. (2002). Urban science education studies: A commitment to equity, social justice and a sense of place. *Studies in Science Education, 38*(1), 1–37 (Online). Available from: https://doi.org/10.1080/03057260208560186.

Bay, J. H. P., & Lehmann, S. (2017). *Growing compact: Urban form, density and sustainability (Online)*. New York, USA: Routledge. Taylor & Francis Group. Available from: https://doi.org/10.4324/9781315563831.

Bourdieu, P. (1986). The forms of capital. In J. Richardson (Ed.), *Handbook of theory and research for the sociology of education* (pp. 241–258). New York, USA: Greenwood Press.

Burton, E. (2000). The potential of the compact city for promoting social equity. In Elizabeth Burton, Mike Jenks, & Katie Williams (Eds.), *Achieving sustainable urban form* (pp. 19–29). New York, US: Routledge.

Cappelletti, A. (2014). *Qiu Xiaolong: così sparisce Shanghai (Online)*. 2014. cinaforum. Available from: http://www.cinaforum.net/intervista-qiu-xiaolong/. Accessed April 10, 2019.

Coleman, J. S. (1988). Social capital in the creation of human capital. *American Journal of Sociology, 94*, S95–S120.

Crossa, V. (2009). Resisting the entrepreneurial city: Street vendors' struggle in Mexico City's Historic Center. *International Journal of Urban and Regional Research, 33*(1), 43–63 (Online). Available from: https://doi.org/10.1111/j.1468-2427.2008.00823.x.

de la Torre, M. I. (2015). Espacio público y colectivo social. *Nova Scientia, 14*(7), 495–510.

Dempsey, N. (2006). *The Influence of the quality of the built environment on social cohesion in English neighbourhoods*. Oxford Brookes University.

Dempsey, N., Bramley, G., Power, S., & Brown, C. (2011). The social dimension of sustainable development: Defining urban social sustainability. In *Sustainable development* (Online) 19 (May 2009), (pp. 289–300). Available from https://doi.org/10.1002/sd.417.

DeWolf, C. (2019). Hong Kong's seat of power, Statue Square—Its colonial history, symbolism, and how domestic helpers made it their own. *South China Morning Post* [Online] 2 February. Available from: https://www.scmp.com/lifestyle/travel-leisure/article/2184584/hong-kongs-seat-power-statue-square-its-colonial-history.

Ekawati, S. A. (2015). Children—Friendly streets as urban playgrounds. In *Procedia—Social and Behavioral Sciences, 179*, 94–108 (Online). Available from: https://doi.org/10.1016/j.sbspro.2015.02.413.

Field, J. (2003). *Social capital*. London, UK: Routledge London.

Fonseca Rodriguez, J. M. (2015). La importancia y la apropiación de los espacios públicos en las ciudades. *Revista de Tecnologia y Sociedad., 4*(7), 1–11.

Gonzalez, M. (2012). Female open space invaders/Ellas filipinas Curator. D. Chipperfield (Ed.), International exhibition Venice Biennale Architecture "Common ground" (Online). Available from: http://www.marisagonzalez.com/home_in.htm.

Graumann, C. F. (1976). *The concept of appropriation (aneignung) and modes of appropriation of space (Online), 6*(2), 301–313. Available from: http://iaps.scix.net/cgi-bin/works/Show?iaps_00_1976_009.

Huxley, A. (1926). *Jesting pilate: An intellectual holiday*. George H. Doran Company.

Kim, A. M. (2013). *Sidewalk City* (Online). Chicago, US: The University of Chicago Press. Available from: https://doi.org/10.7208/chicago/9780226119366.001.0001.

Klienhans, R., Priemus, H., & Engbersen, G. (2007). Understanding social capital in recently restructured urban neighbourhoods: Two case studies in Rotterdam. *Urban Studies, 44*, 1069–1091.

Korosec-Serfaty, P. (1976). Appropriation of space. In *Proceedings of the Strasbourg conference*. IAPC-3. 1976 Strasbourg-Lovaine La Neuve: CIACO.

Kuo, L. (2014). How Hong Kong's "maid trade" is making life worse for domestic workers throughout Asia, Quartz (Online), 19 February. Available from: https://qz.com/176354/how-hong-kong-maid-tradeis-makinglife-worse-for-domestic-workers-throughout-asia/.

Kyle, G. T., Jun, J., & Absher, J. D. (2014). Repositioning identity in conceptualizations of human–place bonding. *Environment and Behavior, 46*(8), 1018–1043 (Online). Available from: https://doi.org/10.1177/0013916513488783.

Lara-Hernandez, J. A., & Melis, A. (2018). Understanding the temporary appropriation in relationship to social sustainability. *Sustainable Cities and Society, 39C*, 1–14 (Online). Available from: https://doi.org/10.1016/j.scs.2018.03.004.

Lara-Hernandez, J. A., Melis, A., & Caputo, S. (2017). Understanding spatial configuration and temporary appropriation of the street in Latin American cities: The case of Mexico City Centre. In H. Boudagh, A. Versaci, F. Trapani, & M. Migliore, et al. (Eds.), *Advances in science, technology & innovation* (pp. 153–173).

Lefebvre, H. (1992). *The production of the space* (1996th ed.). Cambridge, Massachusetts: Blackwell Publishers Ltd.

Li, J. (2014). *Shanghai homes: Palimpsests of private life*. Columbia University Press.

Littig, B., & Griessler, E. (2005). Social sustainability: A catchward between political pragmatism and social theory. *Sustainable Development, 8*(1), 65–79 (Online). Available from: doi:http://dx.doi.org/10.1504/IJSD.2005.007375.

Martínez, E. (2014). Configuración urbana, hábitat y apropiación del espacio. Scripta Nova. *Revista electrónica de geografia y ciencias sociales, XVIII*(493), 20.

McKenzie, S. (2004). *Social sustainability: Towards some definitions* (27) Online. Available from: https://doi.org/10.1016/j.jbiomech.2012.08.047.

Mehan, A., & Soflaei, F. (2017). Social sustainability in urban context. In: *Architectural research addressing societal challenges* (pp. 293–299). CRC Press (Online). Available from: doi:https://doi.org/10.1201/9781315226255-47.

Miquiabas, B. (2015). *After another nightmare surfaces in Hong Kong's Domestic Worker Community, Will anything change? Forbes* (Online). Available from: https://www.forbes.com/sites/bongmiquiabas/2015/03/31/after-another-nightmare-surfaces-in-hong-kongs-domestic-worker-community-will-anything-change/#55b27fff7ee0.

Nasution, I. (2015). Urban appropriation: Creativity in marginalization. *Procedia-Social and Behavioral Sciences, 184*, 4–12.

Pol, E., & Iñiguez, L. (1996). La Apropiacion del Espacio. *Monografías Psicosoaicles Ambientales* 9.

Pol Urrútia, E. (2002). El modelo dual de la apropiación del espacio en Mira, R., Sabucedo, JM y Romay, J. Psicología y Medio Ambiente. Aspectos psicosociales, educativos y metodológicos (pp. 123–132).

Putnam, R. D. (2000). *Bowling alone: American's declining social capital*. New York, USA: Simon & Schuster.

Rapoport, A. (2005). *Culture, architecture, and design*. Chicago: Locke Science Publisher.

Seaman, J., & McLaughlin, S. (2014). The importance of outdoor activity and place attachment to adolescent development in Coös County, New Hampshire. *Building Knowledge for Families and Communities, 208*(37) (Online). Available from: http://scholars.unh.edu/carsey/208/.

Secher, L. K. (2014). *Measuring social sustainability in the small-scale built environment*. Bruxelles: Universite Libre de Bruxelles.

Siu, J. (2018). Why foreign domestic helpers in Hong Kong must live in their employers' home? *South China Morning Post* (Online). 14 February. Available from: https://www.scmp.com/news/hong-kong/community/article/2133377/why-hong-kong-wants-foreign-domestic-helpers-live-their.

Tunas, D. (2008). *The spatial economy in the urban informal settlement*. Netherlands: Internation Forum on Urbanism.

Vidal-Moranta, T., & Pol Urrútia, E. (2005). La apropiación del espacio: Una propuesta teórica para comprender la vinculación entre las personas y los lugares. *Anuario de Psicologia, 36*(3), 281–297 (Online). Available from: https://doi.org/10.1006/jevp.2000.0185.

Wang, A. (2008). *The song of everlasting sorrow: A novel of Shanghai*. Columbia University Press.

Xiaolong, Q. (2013). *Years of red dust: Stories of Shanghai* (Digital ed.). Marsilio Editori: Venezia.

Yory, M. C. (2011). El Concepto de Topofilia entendido como Teoría del Lugar. Revista Barrio Taller (pp. 1–17).

(Temporary) Appropriation (of Space), Makassar, and Urban Kampung

Brendan C. Josey and Diego Ramirez-Lovering

Abstract This chapter will discuss temporary appropriation (TA) in the context of informal settlements and urban *kampung* in Makassar, South Sulawesi, Indonesia. It will explore ways in which instances of TA may be useful in determining approaches for design and implementation of in situ informal settlement upgrading projects by generating a greater contextual understanding of existing urban dynamics and spatial use patterns. It suggests that emergent and regular instances of TA within the informal context are a manifestation of interactions between legal frameworks, socio-spatial histories and practices, community cohesion, changing land occupancy patterns, and settlement morphologies. The chapter advances that TA is an urban phenomenon with potential to be harnessed for design development; however, simultaneously it postulates that the concept of TA should be evaluated and extended by questioning what is understood as appropriation in complex urban contexts. Further to this, the chapter questions what constitutes TA in circumstances where open or public space is non-existent, limited, or contested. It is posited that in such circumstances, TA can serve to assert public space into otherwise contested zones, and is, as such, driven by rhythms of daily practice and routine. In this way, TA can be understood as a process of resistance, manifesting in territories outside of those classified as public and may therefore encompass interstitial/transitional spaces, negotiated spaces (i.e. vacant private land, rights of access), and public infrastructure. Understood as such, TA may be useful (1) for providing a degree of insight into highly context-specific informal settlement and urban *kampung* dynamics; and (2) for understanding limitations of classification and administration of public-space in developing cities, and subsequently (3) for informing design strategies for in situ upgrading projects in fields ranging from public health to infrastructure. Through such an analysis, it is argued that instances of TA may serve to render visible formerly hidden social practices, needs, desires, and shortcomings of existing development patterns, and thus be leveraged and operationalised to inform the design of community engagement practices and built environment solutions.

B. C. Josey (✉) · D. Ramirez-Lovering
Monash University, Melbourne, Australia
e-mail: brendan.josey@monash.edu

A. Melis et al. (eds.), *Temporary Appropriation in Cities*,
https://doi.org/10.1007/978-3-030-32120-8_11

Keywords Makassar · Appropriation of space · Informal settlement · In situ upgrade · Revitalisation

Introduction

Our work analyses and operationalises concepts of *Temporary Appropriation* (TA) in the context of informal settlements, or urban *kampung*, in Makassar, South Sulawesi, Indonesia. We explore ways in which TA may be useful in determining approaches for the design and implementation of in situ informal settlement upgrading projects by generating a greater contextual understanding of existing social dynamics through analysis of spatial use patterns. We observe that patterns of TA within the *kampung* context result from a range of inter-related social dynamics and interactions, legal frameworks, and emerging settlement morphologies, shaped in equal measure by opportunity and need. TA develops through a process of constant negotiation and trial and error, and in this process, leads to important community knowledge generation. This cross-category knowledge resulting from interactions between legal frameworks, socio-spatial histories and practices, community cohesion, changing land occupation patterns, and settlement morphologies can offer important clues to inform the productive transformation of these environments. From this, we propose to reposition TA as an urban phenomenon that reveals important insights into 'local practices' within communities, while simultaneously highlighting limitations and opportunities for action in the existing urban fabric. We suggest that TA can act as a bridge to aid in the translation from theoretical analysis of spatial occupation to its deployment in in situ informal settlement upgrading projects through inclusion in initial project designs, community co-design processes, and on-going engagement frameworks. The social dynamics of TA shape the physical space of informal settlements and transformation processes must account for this influence.

We first provide an overview of the concept of appropriation, positioning it as a relational phenomenon. For the purposes of design and implementation of in situ upgrading projects in informal settlement or urban *kampung* contexts, we evaluate and extend the concept of TA by questioning current understandings and definitions, as well as by examining the physical and non-physical contextual factors that give rise to it. To do so, we provide a contextual analysis of Makassar and its urban *kampung*, followed by an overview of the Revitalising Informal Settlements and their Environments (RISE) program, an interdisciplinary informal settlement revitalisation project that is trialling de-centralised water treatment systems and assessing their impact on human and environmental health. The urban environment represented by communities participating in RISE is undergoing significant development and densification and as such open or public space is often limited and contested. In these contexts, we argue that TA is a tactic used for the creation of a provisional public realm, shaped by rhythms of daily practice and routine. This is followed by an extended conceptualisation of TA as a tactic of resistance and of learning, manifesting in territories outside prescribed or categorised zones, encompassing interstitial/transitional spaces, negotiated spaces (i.e. vacant private land, rights of way), and public infrastructure. Approached as such, TA may be useful in the way it can (1) provide insights into highly context-specific informal settlement and urban *kampung* socio-spatial dynamics; and thereby (2) help researchers, architects, planners, and policy makers alike understand how individuals occupy urban space outside of their prescribed classifications and administrative domains, and subsequently (3) inform design strategies for in situ upgrading projects. This is made apparent in RISE where TA provides insight into existing spatial use patterns that impact design implementation quite directly, as well as into how these patterns may inform the development of community engagement strategies. An analysis of TA also reveals the diversity of liminal spaces, shared spaces, and otherwise overlooked zones and how they are occupied. We argue that TA analysis may render visible formerly hidden social practices, needs, desires, windfalls, and shortcomings of existing development patterns, and thus be leveraged and operationalised to inform design solutions and their attendant participatory engagement frameworks.

Appropriation

Appropriation is a concept historically situated in the fields of philosophy, social-anthropology, and psycho-sociology (de Certeau, 1984; Hegel, 2012; Locke & Laslett, 1960; Marx, 2004; Olivecrona, 1974). Although *appropriation* can be defined loosely as the "taking as one's own or to one's own use", or "the making of a thing private property" (Oxford English Dictionary, 2017), due to its lineage, the concept of *appropriation* reaches past its etymological definition to grapple with psycho-sociological and philosophical understandings of individuals' agency and their relation to the physical and metaphysical world, beyond simplified notions of private ownership or use. These relations may manifest in evolution of cultural practices, or spatial practices whereby *appropriation* is an act of transformation or adaptation of an idea or behaviour, and reference to private ownership is excluded (Frijhoff, 2018: p. 104; Olivecrona, 1974: p. 222). As such, *appropriation* can be understood as a didactic process whereby the individual, through acts of *appropriation,* learns cultural practices and societal functions, and in doing so, determines their relation to a broader social sphere (Graumann, 1976: p. 117; Hajer & Reijndorp, 2001).

Accepting that *appropriation*, as described above, is a relational phenomenon, it is made visible through an individual's relation to their environment. As such, the concept of *appropriation of space* is arrived at through further elaboration of the relational qualities of *appropriation,* the individual, or group, and the territory they occupy. *Appropriation of space* has been adopted as a way of understanding the individual's relation to space, with authors such as Lefebvre (1992), and de Certeau (1984), finding it a useful mechanism via which to frame an individual's relation to, and generation of, their social and physical realms. Lefebvre goes so far as to suggest that the concept of *appropriation* may only be clarified through a critical evaluation of its *relation to space*, positing that a 'natural space' modified to fit the purposes of a specific group may be said to have been appropriated (Lefebvre, 1992: p. 163). He concedes, however, that although instances of appropriated spaces are many, it is very difficult to identify in what respect they are appropriated, how they are appropriated, by whom, and for whom (Lefebvre, 1992: p. 163). For Perla Korosec-Serfaty in her 1973 work, *The Case of Newly Constructed Zones: Freedom, Constraint, and the Appropriation of Spaces,* the concept of *appropriation of space* is intrinsically linked not only to the space as affected by the individual, but also to the individual as he or she is affected by the space; thus she reinforces the communicative, or relational qualities of *appropriation*. Indeed, she postulates that *appropriation of space* has very little to do with ownership, property, or maintenance; rather, she suggests, it is contingent on a set of activities, or 'sequences of daily life'(Korosec-Serfaty, 1973: p. 393). Here, notions of *appropriation of space* intersect with the work ideas of Michel de Certeau, whose seminal 1984 work, *The Practice of Everyday Life,* centres on the tactics employed by groups and individuals to reclaim their agency and autonomy by learning the fabric of the city and its social relations through practices of the everyday, such as walking or engaging in conversation.

For the *appropriation of space* approach to be useful in relation to informal settlement or urban *kampung* upgrading it is important to consider instances of *appropriation of space* as an insight into how cities of high density use urban, rural and peri-urban spaces, regardless of their public/private classification. This is achieved by exposing how these spaces are occupied, appropriated, used, and exploited based on necessities, negotiations, and cultural practices, rather than how they operate as fixed or stable zones that are classified in ways that predicate function (Mehrotra, Vera, Mayoral, Sennett, & Burdett, 2016). By incorporating sequences of everyday activities into an understanding of *appropriation of space*, notions of private ownership or 'intended' use of property or space are muted, and the relational aspects that exist between the individual(s) or the group, and the space they occupy, are privileged (Korosec-Serfaty, 1976: p. 392). The relational quality inherent in *appropriation of space* suggests that further conceptualisations should take into account a broader range of actors or dimensions that operate and exert influence on the individual's experience (Simone, 2011: p. 356); and they may include social frameworks, cultural practices, legal frameworks, governance structures, the built environment, and the individual's learnt relation to all of them. As such, we argue that *appropriation of space* rests on the communication or negotiation between a set of socio-political limitations or expectations, and the individual; and this communication and negotiation may be rendered visible through sequences of everyday activities. With this in mind, the individual's performances within, and his or her relation to, the urban fabric forces existing socio-spatial conditions and legal frameworks to be tested, pushed, and re-evaluated, and thereby generate conditions in which 'multiple dialogues' are made possible within unfixed territories (Mehrotra, 2008: p. 216). The task at hand, then, is to shift focus from how individuals occupy or use space, and begin to untangle how they engage with, manage, and negotiate contested and flexible spaces. These relations are partially revealed through sequences and activities of daily life and have potential to provide insight into existing strategies that serve to "manage and negotiate the layers of large, often neglected parts of any urban society" (Mehrotra et al., 2016: p. 14).

Such relations are described by one citizen of Makassar. Responding to a lack of publicly available open space in their neighbourhood, the resident and their family took it upon themselves to generate a 'public' space on a vacant plot of land adjoining their house. The space was designed by the family and turned into what they describe as a 'house-yard', a 'public' space representing something they "never imagined before… a space with many functions where every citizen may access it anytime, a place where citizens feel comfortable, peaceful, and integrated with other citizens" (Sawedi, 2017: p. 269), predominantly through sequences of daily life, such as prayer, reading, gardening, and exercise. This (temporary) appropriation of the vacant lot is an instance of a social resistance to lack of public space, and its multivalent function offers insight into dynamics of a specific community and serves to illustrate demand for these daily needs to be satisfied. This is demonstrated not through activity itself, but rather by the adoption, stewardship, and protection by the local community of what began as an individual act of appropriation. Such an approach to analysing appropriation of space has potential to inform revitalisation programs such as RISE. To untangle this further, and to illustrate ways in which an understanding of how this proves useful in the deployment of upgrading projects, we will discuss the RISE program, and its urban context, the city of Makassar, in closer detail.

Makassar: Social Context and Public Space

Makassar is the provincial capital of South Sulawesi, Indonesia, an historic port city with approximately 1.5 million inhabitants. The city is a tapestry of mixed-use residential and commercial areas that range from formalised high-rise typologies and high-density urban areas, through to villages or *kampung* of varied formal/informal mixes (Dovey & Kamalipour, 2017). In the past decade, Makassar has seen rapid urbanisation and population growth. A high number of rural-urban economic migrants, coupled with a high degree of commercial and private development, has put strain on the city's land affordability, availability, infrastructure and natural environment, exacerbated by a number of large-scale land reclamation projects along its coastline and riparian zones. Unplanned city growth has resulted in a dramatic decline in the city's public green open space and non-green open space (Bunawardi, Suzuki, & Yuasa, 2016; Nashrah & Si, n.d.; Surya, 2016; Sutherland, 2011). Green open space in Makassar (public and private combined) averages between 6 and 8% of available land despite Indonesian law[1] stipulating a minimum allocation of 10% private green open space and 20% public green open space in cities across the country. The decline in Makassar's public space has led not only to environmental challenges such as increased instances of urban flash-flooding, increased urban heat, and high

[1]Law of The Republic of Indonesia Number 26, Year 2007, Concerning Spatial Management.

air pollution levels (Nashrah & Si, n.d.; Rachman, 2017: p. 17), but also to dramatic change in social behaviours and land-use patterns, increased community tensions due to lack of 'social space', an increase in privatisation of land-use and segregation, and alienation of residents from their cultural practices (Guinness, 2009: p. 69; Kusno, 2000: p. 117). The lack of open public or social space in Makassar has led to streets and roads serving as de facto 'public-spaces' appropriated for social, celebratory, protest, political, and economic activities (Purnamasari, 2012). Indonesian law acknowledges the importance of streets as economic and productive zones, stipulating that national roads, provincial roads, district/city roads, and village roads may be used for activities outside of their intended transit function, for "public and/or personal interests".[2] UNESCO (2017) adds that indeed "connecting spaces, such as sidewalks and streets, are also public spaces", although this is qualified by a caveat, supported by Young (1990: p. 119), insisting that "a public space [is] an area or place that is open and accessible to all peoples, regardless of gender, race, ethnicity, age or socio-economic level" (UNESCO, 2017). This acknowledgement of realities of the multiple and flexible use of roads is positive; however, it does not substitute for the lack of public space in the city, or account for the absence of sidewalks or *accessible* pedestrian spaces in Makassar, which ultimately disqualifies a street or road from classification as a 'public space' under the UNESCO definition.

The plight of public space in Makassar serves to simultaneously highlight and place pressure on the concept of TA (Lara-Hernandez & Melis, 2018; Lara-Hernandez, Melis, & Caputo, 2017; Lara-Hernandez, Melis, & Lehmann, 2019). TA signifies a multifactorial urban phenomenon understood as "the temporary act in which people use public spaces to carry out individual or collective activities other than the purpose that the space was originally designed for" (Fonseca Rodriguez cited in Lara-Hernandez et al., 2019, p. 4) and is said to occur within the bounds of three broad categories relating to: (1) economic activity; (2) leisure activities; and (3) acts of sacralisation" (Lara-Hernandez & Melis, 2018: p. 368). While this understanding of TA accounts for appropriation of specific territories in the urban realm, the parameters of the definition remain unclear, particularly when translating the concept from theory to contexts where space is often contested, such as informal settlements or urban *kampung*. TA, as defined above, suggests a universal understanding of what constitutes 'public space' and presupposes an existing, mutually agreed design intent, along with an intended 'use' of the space that is being analysed, occupied, or appropriated. In order to achieve a better understanding of what constitutes TA in complex urban environments, a further evaluation of what spaces may be 'appropriated', and what is necessitating it, is required. These considerations will take place in the urban *kampung,* the operational context of the RISE program.

[2]Article 127 of the Law of The Republic of Indonesia Number 22, Year 2009, Concerning Traffic and Road Transport.

Urban Kampung: Development and Spatial Relations

Kampung refers to (mostly) low income, off-street neighbourhoods or villages in Indonesia. The word is often used interchangeably with 'urban-village', 'slum', or 'informal settlement'. Such conflations and their associated connotations have served to create a dualised social reality in Indonesia, with a modern 'middle-class ideal culture' clashing with visions of the 'unhygienic' *kampung* (Kusno, 2000: p. 116) (Fig. 1). Although urban *kampung* form an 'inescapable feature' of urban life in Makassar, and Indonesia at large, they are frequently depicted as unsanitary and regressive and a clear distinction is made in society between the middle-class population and the '*rakyut*', or common-people, who occupy urban *kampung* (Guinness, 2009: p. 10). Beginning with Indonesia's New Order regime (1966–1998), prolific and targeted 'developmentalist agendas' have positioned urban *kampung* and their inhabitants on the periphery of national identity.

This positioning of the *kampung* exacerbates an increasing segregation between wealthier and poorer citizens (Guinness, 2009: p. 11; Murti, 2017: p. 333) and coupled with a lack of public space, is leading to greater degrees of tension and intentional social stratification within the country (Guinness, 2009: p. 11), amounting to what is known as 'hyper-individualism syndrome' (Kamil cited in Guinness, 2009: p. 11). It is partly through built expression that *kampung* people resist their relegation to socially peripheral positions. By expressing outwardly 'modern' and 'progressive' housing, some *kampung* residents gradually position themselves in an imagined future in which they too move 'a class upward' (Kusno, 2000: p. 117) (Fig. 2). The combination of social positioning and (often) insecure tenure status, entwined with

Fig. 1 Real estate housing and the 'reality' of the kampung (*Image* Kongres 35 Tahun Pendidikan Sarjana Arsitekture di Indonesia, Jakarta: IAI, 1985 in Kusno 2000)

Fig. 2 Instance of a kampung house incrementally modernising (*Photograph* B. C. Josey)

population influxes and a 'middle-class' architecture that has "[set] itself against the *kampung*" (Kusno, 2000: p. 113), has led to the pursuit of social legitimacy that manifests in a changing built environment that drives increasingly segregated and privatised land-use. Patrick Guinness provides a concise description of *kampung* people and their perceived resilience in his 2009 work *Kampung, Islam and the State of Java:*

> What continues to characterise kampung people's survival strategies is flexible, often forced upon them by economic circumstances. This flexibility pertains to all aspects of life, from rubbish disposal, house construction, healthcare to rites of passage and neighbourhood meetings, leadership patterns, and savings and loan provisions. As part of informal culture, these strategies are always developed in relation to, but also in addition to, formal procedures laid down by outside authorities, and sometimes followed by internal authorities. Guinness (2009: p. 67)

This described self-determination, resilience and inventiveness within the urban *kampung* has led to a tragedy of the commons that is reducing social and open spaces, which is evident in changing built environment patterns in the *kampung.*

Traditionally in Bugis-Makassar culture, the land surrounding a house is a social, familial, and transitional space between the private interior of the house and its public exterior, with the house itself being 'merely for sleeping in' (Murti, 2017: p. 327; Rachman, 2017: p. 5). However, the government-led developmentalist agendas described above have resulted in the not-so-gradual incremental replacement of traditional stilt houses, that allow for transitional and social spaces, with 'more honourable' brick iterations or 'upgrades' (Rachman, 2017: p. 7), which in turn has

Fig. 3 Increased privatisation of kampung housing (*Photograph* B. C. Josey)

driven a greater degree of delineation of private property by fences and house boundaries. This change has lead to the separation, or erasure, of the *halaman rumah* (the house's surrounding land or yard) from the *kampung* (Rachman, 2017: p. 9) (Fig. 3).

In recent years, the house and vehicles have increasingly become the predominant ways of expressing wealth and social status in *kampung*, and the *halaman rumah* and the *ale bola* (under the stilt house), which were once spaces for social, productive, and recreational activities have been reduced, replaced by what are described as "brick civilisations" (Rachman, 2017: p. 9). While the house has traditionally been associated with social status in Bugis-Makassar culture (Hidayat, Amalia, Sakti, Faqih, & Hayati, 2013), what has changed is the symbolism: signs of high status have shifted from *timpa laja* (roof composition) to *bola loppo* (big house), the latter of which has driven increased plot coverage, resulting in reduction of outdoor space, increasing enclosure and individualism (Rachman, 2017: p. 9) (Fig. 4).

This reduction in open space has increased the multivalent use of the street to the point where it serves a vital function within the communities, absorbing a concentration of functions once dispersed throughout the *kampung*. To accommodate this "in the urban *kampung* context, (public) space can be created anywhere and anytime" (Murti, 2017: p. 339), predominantly appearing in "the street and narrow alleyways, the little rivers that split the settlement...the corners, and others, [which can be] turned into [social-space] when needed" (Murti, 2017: p. 339). As such, the street must be negotiated at all times, acting as a primary social, productive, commercial, and celebratory space. So, social-space in urban *kampung* is nebulous and what constitutes 'public space' is elusive; however, regardless of reduced open space

Fig. 4 *Kampung* housing incrementally enclosed, reducing external social space (*Photograph* B. C. Josey, Rachman, 2017: p. 9)

in the *kampung,* the idea of a fluid social space still exists as a relation within the community (Murti, 2017: pp. 338–339). This fluidity of social-space affects revitalisation projects, including RISE and the development of its implementation strategies, which we will now discuss in detail. Prior to a discussion on RISE, and in order to suggest a way to address this unpredictability, the notion of TA needs to be clarified.

Expanding Temporary Appropriation

The case of Makassar and the urban *kampung* is illustrative of the need for an expanded understanding of TA. Frequent and ephemeral occupation of street and interstitial space within and external to the urban *kampung* by vendors, protesters, subsistence farmers, families and playing children, demonstrates that a public realm, too, may be generated ephemerally. This community created public space provides an opportunity for accessible and diverse public encounters which serve to counter the hegemonic use of available open space within the city. These acts of *appropriation* within the ephemeral public realm are permitted by a network of constant negotiations: celebrations close down streets (Fig. 5); traffic slows so a crop might dry in the midday sun (Fig. 6); construction sites store materials on road-sides (Fig. 7); agglomerations of food vendors slow traffic to a crawl, allowing room for people to eat while creating temporary zones in the road safe enough for children to play

Fig. 5 Religious celebration closing thoroughfare in Makassar (*Photograph* B. C. Josey)

Fig. 6 Crops drying on street (*Photograph* B. C. Josey)

(Fig. 8). In contexts where public space is absent, temporary occupation of space for social uses serves as an act of resistance to the lack of existing public-space within the city, and/or as a reclamation or recreation of the public realm. It is an act of cultural assertion, if not of production, that creates conditions "for an ideal public space [in which] people can physically congregate and create and exchange counter-hegemonic dispositions" (Kim, 2012: p. 229).

Fig. 7 Construction material stored on road, traffic bypass (*Photograph* B. C. Josey)

Fig. 8 A cluster of food stalls slows traffic to walking pace, providing space for children and adults alike (*Photograph* B. C. Josey)

From this, we argue Fonseca Rodriguez's (2015) definition of TA does not go far enough to allow for contexts of informal settlements, or urban *kampung*, or contexts where 'public space', or open space is non-existent, highly limited, contested, or is satisfied by a different set of socio-spatial negotiations, occurring in neither public nor private realms. An understanding of TA should not be constrained to occupation of, or activities within, a public space exclusively; rather, TA manifests in amorphous territories—it is effervescent in its relation to the public realm, the private realm and the liminal spaces of the urban fabric. With this in mind, we argue the term *temporary appropriation* (TA) should be expanded to *temporary appropriation of space* (TAS), and should be focussed on a state of becoming, rather than on the occurrence of activities. TAS occurs as (1) a testing ground in which knowledge of the public sphere is generated, challenged, or accepted, whereby the individual learns socio-cultural relations and legal flexibilities as well as negotiable conditions, and/or (2) a result of this learned flexibility, made visible through daily routines, repetition and habit, suggesting an absence, or presence of socio-spatial relations. We posit that TAS is not reliant on categorical definitions of urban spaces as 'public', or 'private'; rather, it appears as a representation of "liminal positions... [and gives] a certain tangibility to the in-between space where hybridisation occurs" (Bhabha, 2012; Hernández, 2010: p. 89). Understood as such, TAS is an assemblage of flexibilities, opportunities, and/or limitations of a given domain, one that is glimpsed through sequences of daily-life—it is a tactic for managing contested, flexible, or shared territories (Fig. 9),

Fig. 9 A demonstration closes down a street in Makassar (*Photograph* B. C. Josey)

Fig. 10 Residents find space to dry their clothes in an urban kampung (*Photograph* B. C. Josey)

a process of learning whereby individuals, groups, families, and the public interact and negotiate use of space from within a milieu of rapid urbanisation (Fig. 10), changing regulations, land-uses, and evolving cultural practices (Fig. 11). This is further explored through the RISE program.

Revitalising Informal Settlements and Their Environments (RISE)

The RISE program is a transnational research project operating in Makassar, and in Suva, Fiji that is measuring the effect of the built and natural environment on human health in a series of informal settlements. An interdisciplinary team comprised of international researchers, practitioners, and local agents is trialling a decentralised water treatment infrastructure that employs a system of constructed wetlands to filter and treat wastewater, while simultaneously aiming to minimise flood risk through improved drainage, and reduce residents' exposure to contaminated water and associated disease vectors. RISE will implement a series of interventions in communities with significant health stressors arising from contact with contaminated water during flood events caused by inadequate sanitation provision. The central component of the intervention is a water-treatment system comprised of a pressure tank, septic tank, sub-surface wetland and a surface wetland (Figs. 12 and 13). This approach to infrastructure allows for decentralised, environmentally sensitive and autonomous water treatment, addressing the inflexibility of 'trunk infrastructure', which is often impractical or too costly to implement in informal settlement conditions. The wetland system moderates wastewater flows and removes pathogens from black-water

Fig. 11 A family and their friends relax on the street, moving when traffic needs to pass (*Photograph* B. C. Josey)

through bio-filtration processes, discharging improved waters into the environment. While the system allows for flexible, narrow, and negotiable configurations, the space occupied is prominent and visible in the physical, and imagined future of participating *kampung*. As such, design iterations and co-design processes, and potential system configurations of the infrastructure must be conceptualised as a negotiation between system parameters and imperatives, and the community members' ideal vision for their limited available spaces, which morph and adapt constantly.

Critical to addressing this is the adoption and stewardship of both the system itself and the implementation process by the participating communities. This challenge is addressed in a number of ways in RISE, including co-design activities and workshops, involvement of community members in key decisions concerning design and delivery of the system, community engagement, and on-going support and engagement post-implementation. The inclusion of community members and participatory methods support the community's development and underpin their understanding of the RISE system while also aiding in the delivery and longevity of the intervention. This said, practical constraints including availability of space, budget, and limited timeframes, necessitate addressing contextual social-spatial complexities and implementation practicalities simultaneously. It is here that a systematic TAS analysis of urban *kampung* can serve to highlight the daily use of space, and also how the community members relate to space available to them. TAS analysis can provide an

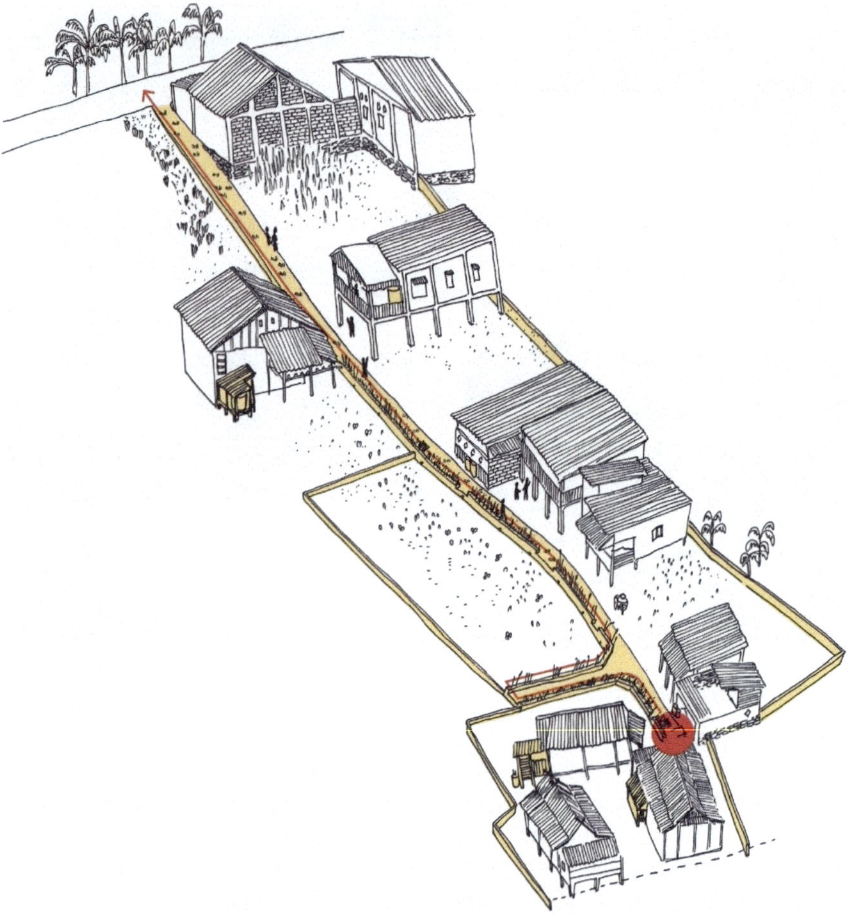

Fig. 12 RISE system diagram (*Diagram* RISE implementation team)

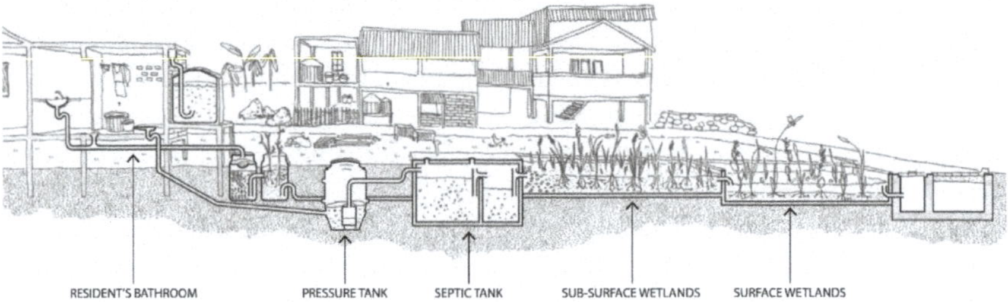

Fig. 13 RISE system diagram (*Diagram* RISE implementation team)

indication of the types of negotiation occurring between individuals and groups in given areas, the adaptation of available space by community members, the variety and frequency of activities within an area, its change over time, and insight into the potential for a space to be utilised for community infrastructure if required. Additionally, TAS can reveal the characteristics of populations appropriating a given space, such as women, children, youth, men, differently abled people, and the elderly. What can be made apparent through TAS is not only that spaces in informal contexts are far from static, but that their functions reveal sets of highly context specific relations that provide insight into the social dynamics of communities and importantly, development patterns influencing community cohesion. Such analysis leading up to, during, and after community engagement processes aids in the design of engagement strategies by providing insight into aspects of community dynamics that impact the implementation of the intervention, and community well-being, directly.

This is illustrated by the incremental development of a small street stall in a *kampung* participating in the RISE program (Figs. 14 and 15). This off-street urban community consists of approximately one hundred and twenty households and is prone to severe flooding. It was the recipient of an upgrading project prior to RISE involvement that consolidated much needed vehicular access into the neighbourhood. The street network in the *kampung* is largely inter-connected and universally used by community members; however, there is a series of cul-de-sacs that have been adopted as quasi-public spaces. A TAS analysis of the *kampung* illustrates ways in which the temporary appropriation of space can morph incrementally into more permanent accretions, revealing characteristics of the settlement. The street stall depicted (Fig. 16) is an instance of community members 'testing' the urban

Fig. 14 Aerial view of a portion of the *kampung*. Highlighted area depicts street (*Image* RISE Implementation Team)

Fig. 15 Street view of incremental accretion (*Photograph* B. C. Josey)

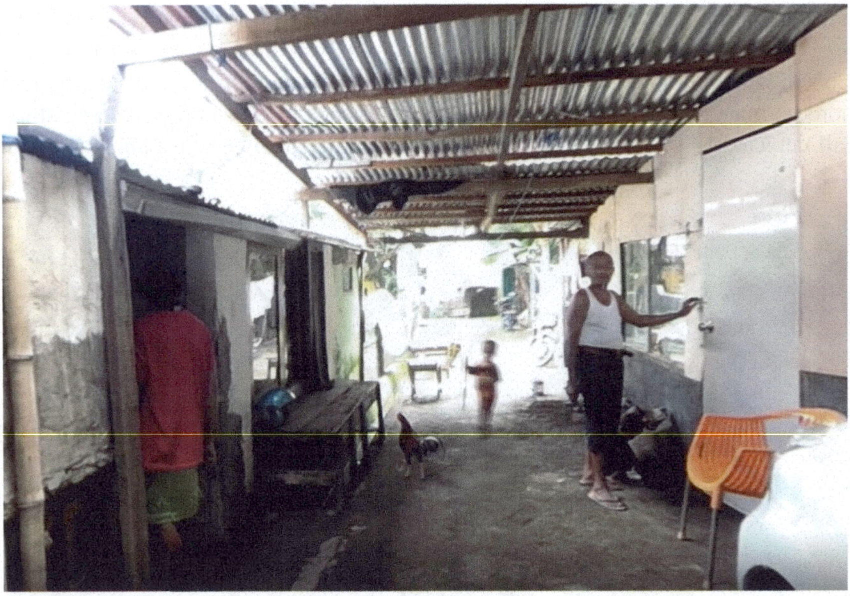

Fig. 16 Appropriation of the street by residents (*Photograph* B. C. Josey)

Fig. 17 More permanent materials replacing structure (*Photograph* B. C. Josey)

open space. As can be observed, the stall began as a temporary structure and is gradually being replaced by more permanent materials (Fig. 17). The structure allows just enough space for a vehicle to pass, maintaining road functionality, while simultaneously appropriating the area through both the physical presence of the structure, and the residents' social behaviour. The space serves as an extension to the existing dwelling that covers the entirety of the residents' land; a roof has been extended over the road, creating a semi-private zone where children play, adults smoke and talk, cars are parked, animals are sheltered, and materials are stored (Fig. 18). This instance indicates that (1) this may not be the first iteration of the stall, (2) the street is predominantly inhabited by a single social or family group, (3) the road is not utilised by vehicles as regularly as other thoroughfares, (4) the small stall may be servicing mosque-goers who use the street as a walking path to reach the mosque at the end of the cul-de-sac, and (5) the social group responsible for the extensions and accretions over the public road have a degree of stewardship over the area. This information, and similar instances, can be incorporated into the design of engagement activities and co-ordination of intervention strategies through clustering of service designs and maintenance strategies. For example, the RISE system is organised around the capacity of each pressure tank and wetland system. Insight into community groupings aids the co-ordination of each component, and understanding such dynamics prior to co-design processes and system configurations allows for the project to anticipate possible outcomes and limitations that may occur throughout the process, including land allocation and maintenance strategies. This analysis and observation can extend beyond the scale of the participating communities and include a variety of urban dynamics, so as to provide insight into how individuals or communities perceive,

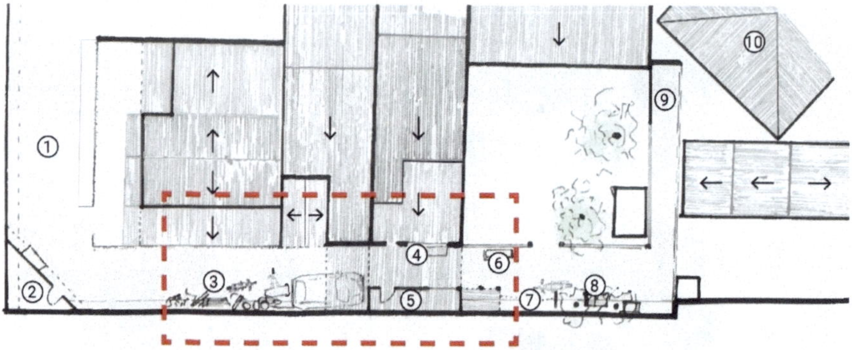

1. PUBLIC ROAD
2. GUARD POST
3. MATERIAL STORE AND BIKE PARKING
4. SEATING AND SOCIAL AREA
5. STREET STALL EXTENDING OVER ROAD AND DRAIN
6. SEATING AND SOCIAL AREA
7. CLOTHES LINE
8. CHICKEN COOPS
9. ACCESS PATH
10. MOSQUE

Fig. 18 Temporary appropriation of space leading to permanent accretions (*Diagram* B. C. Josey)

relate to, and appropriate the various urban spaces available to them, both in the context of his or her community and in their daily lives. Approaching the potentially overwhelming unpredictability of urban *kampung* through a lens focussed on daily use-patterns or practice reveals that firstly, space can be created 'anytime' based on negotiations and compromise; secondly, that cultural or local practices (as related to urban fabric) have a degree of flexibility and adaptability; and thirdly, that activities are rendered visible in the space that remains.

Conclusion

Embedded in *appropriation of space* are dynamics of communication and negotiation from which emerge acts of learning social, cultural, and legal limitations specific to the territory that is being appropriated. The conditions (cultural practices, socio-legal constraints or boundaries) surrounding these territories vary, as do the negotiations that occur within them. It is the *appropriation of space* that exposes (through interactions between the built environment and individual(s) or groups) the slivers of social relations, legal frameworks, cultural practices, limitations and flexibilities that exist within the assemblage of social, political, and legal structures that manage a state, region, or neighbourhood. This is not to suggest that the instances of *appropriation of space* that generate the ephemeral public domain substitute inadequate planning policy or measures; rather, they serve to highlight ways in which a relational approach to *appropriation of space* may serve to render visible civic discourses otherwise hidden or obscured. Such instances also inform as to how informal settlement upgrading strategies are arrived at, how policy makers approach legislation, and contribute to strategies conceptualising urban spaces in complex urban environments. This said, care must be taken to avoid over prescribing value to observations, and TAS should be used in conjunction with a suite of other methods of design analysis, and community engagement.

This chapter has served to expand the notion of TA to TAS through emphasising the relational characteristics of *appropriation*, suggesting that TAS may prove to be an effective and flexible analytical tool for understanding ways in which 'local knowledge' is generated and deployed as a tactic for managing socio-spatial stressors in informal settlement contexts. This understanding of TAS will serve to aid in the production of a more textured, contextualized understanding of the way space is perceived and related to by individuals or groups. More work needs to be done in determining suitable methods with which to apply TAS in the varied contexts in which informal settlement upgrading occurs. TAS has great potential to operate not only as a tool to generate more nuanced understandings of *kampung* community members' relation to space, but also as a participatory method in and of itself, whereby co-design processes and implementation strategies are developed simultaneously from the outset of a project, potentially involving marginalised community members such as women, youth, and children. Approached in this way, TAS's power as a component in democratising design and implementation of informal settlement upgrading projects may be amplified.

References

Bhabha, H. K. (2012). *The location of culture*. London, UK: Routledge.

Bunawardi, R. S., Suzuki, Y., & Yuasa, H. (2016). Diversity and utilization of public space in Rusunawa Mariso, Makassar-Indonesia. *Journal of Asian Architecture and Building Engineering., 15*(3), 433–440.

de Certeau, M. (1984). The practice of every day life. In S. Rendall (Ed.). Los Angeles, USA: University of California Press.

Dovey, K., & Kamalipour, H. (2017). Informal/formal morphologies. In: *Mapping urbanities* (pp. 223–248). London: Routledge.

Fonseca Rodriguez, J. M. (2015). La importancia y la apropiación de los espacios públicos en las ciudades. *Revista de Tecnologia y Sociedad, 4*(7), 1–11.

Frijhoff, W. (2018). Foucault reformed by Certeau: historical strategies of discipline and everyday tactics of appropriation. In *Cultural history after Foucault* (pp. 83–100). London: Routledge.

Graumann, C. F. (1976). The concept of appropriation (aneignung) and modes of appropriation of space. [Online]. *6*(2), 301–313. Available from: http://iaps.scix.net/cgi-bin/works/Show?iaps_00_1976_009.

Guinness, P. (2009). *Kampung, Islam and state in urban Java*. Singapore: NUS Press.

Hajer, M. A., & Reijndorp, A. (2001). *In search of new public domain-analysis and strategy*. Rotterdam NAI: Rotterdam.

Hegel, G. W. F. (2012). *The phenomenology of mind*. Courier Corporation.

Hernández, F. (2010). *Bhabha for architects*. London: Routledge.

Hidayat, R., Amalia, L., Sakti, M., Faqih, M., & Hayati, A. (2013). The expression of social status on house form; Case: Bugis-Makassar modern house. *International Journal of Scientific and Engineering Research, 4*, 3.

Kim, A. M. (2012). The mixed-use sidewalk: Vending and property rights in public space. *Journal of the American Planning Association., 78*(3), 225–238.
Korosec-Serfaty, P. (1973). The case of newly constructed zones: Freedom, constraint and the appropriation of spaces. *Architectural Psychology*, 389–396.
Korosec-Serfaty, P. (1976). Appropriation of space. In: Proceedings of the Strasbourg Conference (IAPC-3). Strasbourg-Lovaine La Neuve: CIACO.
Kusno, A. (2000). *Behind the postcolonial: Architecture, urban space and political cultures in Indonesia*. New York, USA: Routledge.
Lara-Hernandez, J. A. & Melis, A. (2018). Understanding the temporary appropriation in relationship to social sustainability. *Sustainable Cities and Society*. [Online] *39C*, 1–14. Available from: https://doi.org/10.1016/j.scs.2018.03.004.
Lara-Hernandez, J. A., Melis, A., & Caputo, S. (2017). Understanding spatial configuration and temporary appropriation of the Street in Latin American cities: The case of Mexico City Centre. In H. Boudagh, A. Versaci, A. Sotoca, F. Trapani, M. Migliore, et al. (Eds.), *Advances in science, technology & innovation* (pp. 153–173).
Lara-Hernandez, J. A., Melis, A., & Lehmann, S. (2019). Temporary appropriation of public space as an emergence assemblage for the future urban landscape: The case of Mexico City. *Future Cities and Environment*. [Online] *5*(1), 1–22. Available from: https://doi.org/10.5334/fce.53.
Lefebvre, H. (1992). *The production of the space* (1996th ed.). Cambridge, MA: Blackwell Publishers Ltd.
Locke, J., & Laslett, P. (1960). *Two treatises of government: A critical edition with an introduction and apparatus criticus*. Cambridge: Cambridge University Press.
Marx, K. (2004). *Capital: A critique of political economy*. London, UK: Penguin.
Mehrotra, R. (2008). Negotiating the static and kinetic cities: The emergent urbanism of Mumbai. In *Other cities, other worlds: Urban imaginaries in a globalizing age* (pp. 205–218).
Mehrotra, R., Vera, F., Mayoral, J., Sennett, R., & Burdett, R. (Eds.). (2016). Ephemeral urbanism : Cities in constant flux. Santiago, Chile: ARQ ediciones (Primera ed.).
Murti, Y. (2017) Urbanism of Yard. In: A. J. Rachma (Eds.), *Halaman Rumah/Yard* (pp. 323–340). Makassar: Tanahindie Press.
Nashrah, S. T. & Si, M. (n.d.). Optimalisasi Distribusi Ruang Terbuka Hijau di Kelurahan Banta-Bantaeng Kecamatan Rappocini Kota Makassar.
Olivecrona, K. (1974). Locke's theory of appropriation. *The Philosophical Quarterly, 24*(96), 220–234.
Oxford English Dictionary. (2017). *Appropriation*, n. [Online]. 2017. Oxford University Press. Available from: http://www.oed.com/view/Entry/9877?redirectedFrom=appropriation. Accessed November 3, 2017.
Purnamasari, A. (2012). Kajian Spasial Ruang Publik (Public Space) Perkotaan Untuk Aktivitas Demonstrasi Mahasiswa di Kota Makassar. *Jurnal Bumi Indonesia,* 1(2).
Rachman, A. J. (2017). We are thriving in the city, we are flourishing in the yard. In A. J. Rachman (Ed.), *Halaman Rumah/Yard* (pp. 1–23). Makassar: Tanahindie Press.
Sawedi, M. (2017). Home yard, public space, and happy city: An imagination of a common resident. In A. J. Rachman (Ed.), *Halaman Rumah/Yard* (pp. 251–271). Makassar: Tanahindie Press.
Simone, A. (2011). The surfacing of urban life: A response to Colin McFarlane and Neil Brenner, David Madden and David Wachsmuth. *City, 15*(3–4), 355–364.
Surya, B. (2016). The processes analysis of urbanization, spatial articulation, social change and social capital difference in the dynamics of new town development in the fringe area of Makassar City (case study. In Metro Tanjung Bunga Area, Makassar City). *Procedia-Social and Behavioral Sciences, 227*, 216–231.
Sutherland, H. (2011). Whose Makassar? Claiming space in a segmented city. *Comparative Studies in Society and History, 53*(4), 791–826.

UNESCO. (2017). Inclusion through access to public space. United Nations Educational, Scientific and Cultural Organization. [Online]. 2017. Social and Human Sciences. Available from: http://www.unesco.org/new/en/social-and-human-sciences/themes/urban-development/migrants-inclusion-in-cities/good-practices/inclusion-through-access-to-public-space/. Accessed: March 27, 2019.

Young, I. M. (1990). *Justice and the politics of difference*. Princeton: Princeton University Press.

Extending Temporary Appropriation Through Architecture: The Role of Adaptive Reuse in Shaping New Zealand's Built Environment

Paola Boarin

Abstract The chapter explores the multiple meanings and implications of integrated building adaptation and its role in the retention and re-creation of heritage values and sense of place. The interdependence between built and natural environment, as well as the building and urban dimension, is discussed to highlight its relationship with adaptive reuse and temporary appropriation. These concepts are further analysed to appreciate how New Zealand's cultural plurality enriches the understanding of cultural heritage and the approaches related to the adaptation of the built environment. If, on one side, New Zealand was built through architectural concepts and urban practices strongly inspired by Western models, on the other side, the country is now experiencing the emergence of new architectural languages, which are becoming more relevant in addressing existing buildings and historic heritage. In this regard, the contribution discusses how contemporary architecture in New Zealand is conveying cultural, environmental and social values, especially in the adaptive reuse of existing and historic buildings, in the attempt of establishing a new dialogue between tradition and innovation that could overcome temporality and extend the concept of appropriation of space.

Keywords Historic buildings · Adaptive reuse · Temporary appropriation · New Zealand architecture

The Impacts of New Uses in the Adaptation of Historically Significant Buildings

The topic of conservation of historically significant buildings through new uses is of great interest among scholars and practitioners as architectural heritage is an expression of our history and contributes to the understanding of the past's relevance in regards to our contemporary life and civilization. We are today perfectly conscious of how much the loss of the original purpose of a building can cause its rapid decay, but we are also aware that any act of building conservation without the correct choice

P. Boarin (✉)
School of Architecture and Planning, University of Auckland, Auckland, New Zealand
e-mail: p.boarin@auckland.ac.nz

A. Melis et al. (eds.), *Temporary Appropriation in Cities*,
https://doi.org/10.1007/978-3-030-32120-8_12

of new uses can lead to greater dangers for our cities and society. In this regard, the 'integrated conservation' approach declared in the 1975 European Charter of the Architectural Heritage, which encourages the development of initiatives characterised by a positive synthesis between "sensitive restoration techniques and the correct choice of appropriate functions" (Committee of Ministers of the Council of Europe, 1975: p. 3), is still largely applicable to planning and design initiatives nowadays. The role of the architect and that of design become then paramount for achieving a balance between the old and the new, through a rigorous history research and the development of an informed creative process.

Building adaptation has been investigated extensively (Plevoets, Van Cleempoel, & WIT Transactions on The Built Environment, 2011). Douglas (2006): p. 1 defines adaptation as "any work to a building over and above maintenance to change its capacity, function or performance". The Burra Charter defines adaptation as "changing a place to suit the existing use or a proposed use" (Australia ICOMOS, 2013: p. 2), focussing on the concept of 'place' (i.e. a geographically defined area including natural and built environments with tangible and intangible value) rather than on the physical boundary of a building. This association was supported by ICOMOS New Zealand as well in their Charter for the Conservation of Places of Cultural Heritage Value, where adaptation is defined as "the process(es) of modifying a place for a compatible use while retaining its cultural heritage value" (ICOMOS New Zealand, 2010: p. 9), pointing out that this process includes both alteration and addition activities.

The scope of adaptation works can be very wide and depends on the extent and nature of the changes proposed for the building in order to respond to the transformed needs of a society (Douglas, 2006) (Fig. 1), but it also depends on the shearing layer of change considered. As discussed by Brand (1994), a building is composed by multiple dimensions progressively moving from the inner core (the people inhabiting

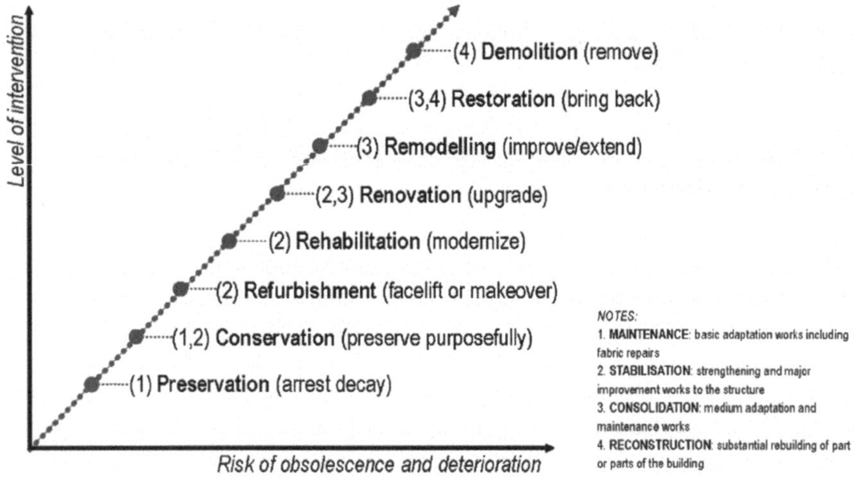

Fig. 1 Levels of adaptation (Author's graphic elaboration based on Douglas, 2006, p. 3)

it) to the outside (furniture and equipment, space plan, services, building structure, building skin) and towards the surrounding site (location and context). The more we move toward the outside and the surrounding, the higher the lifespan of the layer considered. Therefore, the impact of adaptation works is not limited to the physical boundary of the building, but it extends to the environment beyond that to enter the realm of urban design, defining a stronger connection with the city and the wider community.

If the choice of compatible new uses for old structures does not depend on and affects the building level only, but must also find a response at urban level to fulfil its social and environmental purpose, the adaptation project should also consider the well-being of people outside the building, in order to create a multi-faceted and efficient setting where structures are completely or partially accessible by the public. In line with this approach, as the integrated requalification of a building and its surroundings creates new dynamics and increases the sense of belonging of a local community towards the urban environment, the adaptation project should encourage the integration of a variety of uses within the same initiative, as this will improve the fruition of the structure and support cultural diversity. The mutual dependence between the building and the urban scale and the role played by the adaptive reuse process in the re-establishment and valorisation of their role for the society outlines a connection with the concept of temporary appropriation as defined by Fonseca Rodríguez (2015) as "the temporary act in which people use public spaces to carry out individual or collective activities other than the purpose that the space was originally designed for". The relationship between reuse and appropriation is strong, as some of the requirements needed for allowing the built environment to adapt to new uses can be extended to public spaces in order to enable their temporary appropriation. Among others, these are convertibility, spatial and technological flexibility and expandability, attributes that can define both the building's vocation to change to fulfil new purposes and the public space's vocation to enable different uses and an occupation by different people. Also, the spatial, architectural and technological dimension of the adaptation approach in regards to the host building (Boarin et al. 2016; Wong, 2017) can generate different responses in terms of appropriation of space.

Understanding the Links Between Building Adaptation and Urban Design in New Zealand

The New Zealand's built environment that we know today is the result of the urbanisation that followed the colonisation of the country in the mid-nineteenth century. Town planning was the result of private corporations' initiatives, commercial imperatives and local and central government's legislative measures and, as such, the urban pattern of colonial cities in New Zealand was shaped according to grid layouts similar to those used before in the colonisation of the USA and Australia (New Zealand Productivity Commission, 2015: p. 1). This grid model was perhaps functional and efficient in a commercial, developmental and safety point of view, but did not consider the existing environmental morphology (as the shape of the land) and social dynamics, neither the existing ones of the indigenous population nor potential future ones. In this context, it is very meaningful to mention that, being a concept born in the Western world, the term 'urban design' did not (and still doesn't) resonate with the indigenous Māori population's worldview. From the Māori perspective, urban design is a reductive approach, which does not fulfil the understanding and acknowledgment of the complex relationships between the multiple interconnected dimensions of the world we live in and, consequently, how we address its transformations.

> As Māori we have a unique sense of our 'landscape'.
>
> It includes past, present and future.
>
> It includes both physical and spiritual dimensions.
>
> It is how we express ourselves in our environment.
>
> It connects whanau and whenua, flora and fauna through whakapapa.
>
> It does not disconnect urban from rural.
>
> It transcends the boundaries of 'land'scape into other 'scapes'; rivers, lakes, ocean and sky. […]
>
> It is not just where we live – it is who we are! Te Aranga (2008: p. 3)

This definition of 'Māori cultural landscape', which was developed for the Auckland Design Manual (Auckland Council, 2013) as a response to the lack of consultation with Māori groups in regards to the 2005 Urban Design Protocol, overcomes that of urban design and gets closer to the concept of 'historic monument' as defined by the Venice Charter (ICOMOS, 1964: p. 1), to the combined concepts of 'cultural heritage' and 'natural heritage' (United Nations Educational Scientific and Cultural Organization, 1972: pp. 2–3) and the concept of 'places of cultural heritage value' (ICOMOS New Zealand, 2010). What is common across these definitions is that they encompass both tangible and intangible values and they acknowledge the linkages between built and natural environment, with the purpose of retaining and exposing those values and supporting their functions to the benefits of current and future generations, which is the ultimate goal of conservation.

To recognise the unique Māori identity and the distinctive cultural environment of Aotearoa/New Zealand in the design process, the Auckland Design Manual (a practical guide that sits alongside the Unitary Plan, i.e. Auckland Council's regulatory land-use planning document), has integrated the Te Aranga Māori Design Principles, which outline the process for positively engaging with Mana Whenua (the Authority over land and natural resources) and shape the built environment. The Te Aranga Māori Design Principles were inspired by wider Māori values, among which is worth mentioning the following (Auckland Council, 2013):

- *Kaitiakitanga*, i.e. "managing and conserving the environment as part of a reciprocal relationship, based on the Māori world view that we as humans are part of the natural world";
- *Wairuatanga*, i.e. "the immutable spiritual connection between people and their environments";

- *Whanaungatanga*, i.e. "a relationship through shared experiences and working together which provides people with a sense of belonging".

Bringing and translating these principles into the adaptation of the built environment, whether this involves a single building or the wider urban context, means to incorporate those values of guardianship and stewardship that are crucial to the fulfilment of the act of valorisation of the built heritage, through a collaborative and integrated effort that brings multiple stakeholders (private sector, government and wider society) and their needs together. As adaptive reuse follows these principles and values, so does temporary appropriation.

Heritage, Identity and the Appropriation of Space

The protection and enhancement of historically significant buildings and places is generally taken for granted in the European environment, but becomes of paramount importance in those countries with a more recent building stock, such as New Zealand. If cultural heritage is "the legacy of physical artefacts and intangible attributes of a group or society that are inherited from past generations, maintained in the present and bestowed for the benefit of future generations" (United Nations Educational Scientific and Cultural Organization, 2017), this heritage can be found in 'younger' cities too and must be safeguarded through the most appropriate acts of adaptation. Unfortunately, the argument of the lack of heritage value has been used very often for demolishing historically relevant buildings in cities with a recent past in order to make space for new commercially driven initiatives. Provided that an heritage assessment performed by experts prior to starting any initiative is always essential for understanding the qualities and values of the buildings and their future capacity, maintaining the existing historic environments by means of adaptation can become an even more powerful driver for the rehabilitation and redevelopment of entire urban areas, as discussed in the so-called Declaration of Toledo (EU Member States Ministers, 2010).

> Revaluing deteriorated public spaces and providing new open spaces, together will the protection or requalification of architectural forms, will contribute not only towards the improvement of the urban scene, landscape and place quality of many of our cities' urban fabrics, and therefore to raise their attractiveness to highly skilled workers and business, but also to increase their attractiveness and the local residents' identification with the urban environment and their community. This will therefore contribute not only to their cultural enrichment but also towards creating or recreating citizenship, because the values of democracy, coexistence, exchange, civic progress, diversity, living together and freedom are key factors in the culture of the European city, which are expressed most effectively in the public realm. EU Member States Ministers (2010)

If heritage has the power of redefining a location (Wilkinson, Remøy, & Langston, 2014: p. 176), rehabilitated heritage can create new centres of social interaction, by changing the perception of the building itself and that of the spaces and streets adjacent to it, reverting biases against deteriorated environments, in line with Newman's idea of 'defensible space' (Newman, 1973), and with the positive effect of triggering further developments in the same area. The symbiotic relationship between a building and its location is therefore maintained and enhanced through the protection and adaptation of historic buildings, contributing to the notion of place making where multiple socio-cultural identities can thrive. The idea of identity is also essential for understanding the connection between adaptive reuse and appropriation of space. According to Filep, Thompson-Fawcett, and Rae (2014), identity is defined by the ways in which human beings find their place in the world in regards to a multiplicity of factors (such as culture, space, economy, religion, etc.) and can be described by three aspects, i.e. well-being, plurality (and diversity) and inclusivity. In the case of adaptive reuse, the process is activated by the place being still recognised as the manifestation of a society's cultural expression through its craftsmanship and, therefore, as a demonstration of its identity. In the case of temporary appropriation, people tend to occupy a space when they feel identified or represented by it and this contributes to group cohesion (Lara-Hernandez & Melis, 2018). Hence, finding new uses for heritage landscapes, by means of an integrated process of building adaptation, urban regeneration and temporary appropriation, contributes to the creation of social value and cultural identity. This cannot be attributed to just a single building, but the way one building relates to another, to the ensemble and to the public space in between is what creates the identity of a place (English Heritage, 2009). This was particularly important for the Auckland's Britomart precinct, where the conservation and enhancement of its cultural significance and heritage values through the restoration and adaptive reuse of many of its peripheral buildings became a driver for a variety of developments in the area.

The Britomart precinct is the largest heritage-regeneration project in New Zealand and is today a vital and very dynamic part of the city. The land was once owned by the Māori tribe Ngāti Whātua who donated it to Governor Hobson for building a new capital city and most of the Britomart area, as we know it today, was reclaimed from the sea in the 1870s, with the aim of building port facilities. In the late nineteenth century, the Britomart became Auckland's main business district, comprising a railway station, a large number of warehouses and trading premises and governmental offices, but this central role started to decline in the first half of the twentieth century and the precinct suffered from lack of investments and increasing decay that led the area to the abandonment. During those decades, New Zealand Historic Places Trust managed to attribute heritage protection to the buildings defining the precinct's area and this forward-looking strategy prevented these structures from being demolished. In the early 2000, Auckland City approved a masterplan for the development of a downtown underground rail station, a transport interchange centred on the historic Central Post Office and the revitalisation of the built heritage of the Britomart precinct. The underground station allowed most of the surface area to be pedestrianised and the conservation and enhancement of the protected 18 heritage buildings, mostly Victorian warehouses, recovered the character of the place and gave scale and enclosure to the precinct. This heritage-led regeneration eventually created a strong sense of place given by both the presence of a large number of historic buildings, which were adaptively reused and upgraded to guarantee their

Fig. 2 The Britomart precinct in Auckland in different types of temporary appropriation of the open space, from farmers' markets (left), to spaces for people to meet (right) and public events (centre), which make the area equally vibrant by day and by night. Photos: Paola Boarin (left), https://www.thedenizen.co.nz/culture/cinema-peroni-2019/ (centre), https://www.localist.co.nz/l/britomart (right)

inhabitability, and by the new possibilities for the community to occupy the open space between the buildings through a variety of activities. The temporary appropriation of this space takes the multiple forms of activities related to economy (farmers' markets and other forms of trade) and leisure (sports, artistic expressions, etc.), as defined by (Lara-Hernandez & Melis, 2018) (Fig. 2). The relationship between the retention of heritage values of the buildings, the creation of an open yet protected environment and the multiplicity and flexibility of its uses give the Britomart the cultural distinctiveness and social dimension that we recognise today and that has become a model for many other regeneration activities across New Zealand.

Permanence and Impermanence in the Creation of New Public and Shared Spaces

If it is argued that appropriation allows people to produce urban space (and places) rather than just to use already existing urban space (Purcell, 2002), it can also be said that design (architectural and urban) plays a pivotal role in defining new or recovered relationships between society and built environment. A successful adaptation project can facilitate the creation of new places, in the building and around it, that are resilient and future-proofing, thus enabling an ever-evolving appropriation of a space that changes with time and human needs, but co-exists with the physical environment in a positive relationship of mutual support, rather than what, in the case of appropriation, is often understood as a parasitical interference.

Although there are many successful adaptive reuse projects in New Zealand and worldwide, the role of the surrounding environment seems to be often underestimated and neglected in such projects. A good design for the building is crucial to the entire operation, but its connections to the existing urban fabric can make a distinction between an emerging, yet isolated, architectural gesture and an act of cohesion between a new or re-discovered identity for a historically significant object and the society that will benefit from it. A careful planning and design of the spaces surrounding the building will therefore enrich the user's experience, as well as extend the life of the building itself even further by becoming a place where multiple dimensions are blended together and coexist.

An interesting example where the design for adaptive reuse extends its impact to the urban surrounding is the project for the Auckland Art Gallery Toi o Tāmaki, the result of a collaboration between the Australian designers Francis-Jones Morehen Thorp (FJMT) and the New Zealand designers Archimedia. The project, completed in 2011, includes the restoration and adaption of the heritage buildings located in the heart of the city, a new building extension and the redesign of adjacent areas of Albert Park, Auckland's central gardens. The new extension of the 1988 building replaces a previous extension added in the 1970s, which was considered dysfunctional to host both the art collection and the new additional activities open to the public (such as education, child and family spaces), as well as to achieve the aspirational goals of the city to find a rightful place in the international arts scenario.

Even if the impacts of the project on Albert Park and the surrounding streets were long discussed in the Environment Court over a three-years protracted process, the engagement with the park defines one of the most successful aspects of this new insertion in regards to temporary appropriation of the urban space. The architecture language is driven by a biomimicry process informed by the surrounding landscape and natural environment, where a series of tree-like canopies, inspired by the adjacent Pōhutukawa trees and assembled from carefully selected Kauri from recycled sources, shape and cover the entry forecourt, atrium and gallery areas and extend to the back towards the slope of the park (Fig. 3). The transparency of the façades enclosing the volume underneath the canopies allows a complete view through the gallery spaces towards the green landscape, generating an escalating visual continuum from the open place in front of the Art Gallery to Albert Park. Here, an open-air paved space covered by additional canopies is created as a *trait d'union* between

Fig. 3 The Auckland Art Gallery Toi o Tāmaki, recipient of several national and international awards, among which the 2013 World Building of the Year Award. The project involves the restoration and adaptation of the historic building designed by architects John H. Grainger and Charles A. D'Ebro in 1888 in a 'French Château' style and the design of a new extension (image on the left hand side). The extension has provided the building with additional indoor and outdoor flexible spaces that unite the building with Albert Park on the back (image in the centre) and to the rest of the city to the front (image on the right hand side), maintaining a visual connection through glazed facades and the use of biomimicry architectural strategies for the wooden canopies. Photos: Paola Boarin

the built and natural environment and is open to the community who can use it in multiple ways. Observational activities confirmed that this newly generated public space is consistently used through temporary appropriation, mainly ascribable to the leisure category as defined in Lara-Hernandez, Melis, & Caputo (2017). The temporary appropriation pattern is consciously and carefully enabled by the adaptive reuse of the historic building rather than being an unplanned and uncontrolled outcome and this has a positive impact on how this new public space is perceived and used by people. Several spontaneous and community or individuals-driven activities take place here: from dance and music rehearsals to school trips' outdoor educational activities, from sports-related exercises to simple resting or aggregation activities. These are completely amalgamated with the everyday flows of people using the connection path tangential to the open space, which, in urban terms, becomes both a pole and a node. The various uses of the space defining its temporary appropriation appear as an extension of the cultural environments enabled inside the Art Gallery and this is due to the design action, which successfully shaped the open space within the context of a defined adaptation programme, while intentionally leaving its ultimate purpose to the decision of the community.

If it has been observed that temporary appropriation does not happen in any space, but only in those where people develop a sense of belonging and a cultural bond (Madanipour, 1996; Relph, 1976), the experience of the Auckland Art Gallery shows how the architectural qualities of the new built environment (spatial flexibility as first) and the newly established relationships with nature can foster the generation of new cultural landscapes within the established urban fabric.

If the positive effects on cultural and on environmental sustainability produced by the adaptive reuse of historically significant buildings are well known today (Conejos, Langston, & Smith, 2011; Misirlisoy & Günçe, 2016; Plevoets & Sowińska-Heim, 2018; Wilkinson et al., 2014; Yung & Chan, 2012), but the effects on social sustainability, especially in conjunction with the concept of temporary appropriation, are less discussed. Adaptive reuse responds to the necessity of adapting a building to changed human needs, while temporary appropriation responds to the necessity of adapting human needs to a changed built environment, which can be a building or a wider urban context. Both respond to human needs and depend on temporality in terms of permanence and impermanence. In this sense, the act of adaptation and any initiative supporting and enabling ever-evolving human needs through the built environment become powerful means to enable a sustainable development in the broadest sense of "development that meets the needs of the present without compromising the ability of future generations to meet their own needs" (World Commission on Environment and Development, 1987: Chap. 2).

The effort in uniting all the three realms of social, environmental and economic sustainability, underpinned by the definition of sustainable development, under a coherent and comprehensive design activity is demonstrated in the Mason Bros building in Auckland designed by the architectural firm Warren and Mahoney. The project involves the adaptive reuse of a 1920s character warehouse building into a commercial development located in the Wynyard Quarter Innovation Precinct, one of the most extensive, challenging and innovative developments in the Auckland's waterfront. The adaptive reuse is very successful in finding a balance between the transformation and conservation of the existing urban fabric by creating a strong formal and material contrast between the old and the new, pursuing a "works *with* the building stock" approach rather than a "works *in* the building stock" one (Giebeler & Kahlfeldt, 2009: p. 18). A highly reflective glazed object is suspended within the existing volume, leaving a separation gap from the surrounding vertical walls on three sides, thus enhancing the existing surfaces, which are voluntarily left uncovered to show the irregularities of the original bricks and concrete structures. This "rough style" approach (Kramer, 2015), preserves the evidence of the past telling about a previous life of the building, while enabling a new way of living the space responding to contemporary needs (Boarin et al., 2016).

In the adaptation project, private and public spaces are combined together into the 60-m long internal space, which serves multiple purposes: it defines the building's main circulatory system, it provides an internal pedestrian link between the two sides of the precinct (thus also allowing a direct fruition of the heritage building by the public) and it creates a flexible open plan covered space suitable for a number of spatial modifications (Fig. 4). This space is simultaneously a street and a plaza, a public and a private space, and, as such, it enables multiple activities related to temporary appropriation. Here, staff working in the building meet and collaborate, people not working in the building find a place for interaction with other people, external stakeholders find a place for exhibitions, talks and aggregation, etc., all enabled by an ever-evolving and fluid space that activates and engages with the external urban fabric and, at the same time, it represents its extension. The limited range of materials selected for the adaptation (steel, concrete, glass and timber) is an integral part of the strategy, creating a strong connection with the industrial legacy and heritage character of the building while enabling the atrium/lane to be heavily inhabited through temporary activities.

The Mason Bros project combines the programme (use), the technique (performance) and architecture (language) together, enhancing and highlighting the relationship between a past life, which is shown in all its evolutions and imperfections,

Fig. 4 The Mason Bros Building in the Auckland Wynyard Precinct is an adaptive reuse of a character industrial warehouse into a multi-tenancy commercial development, which received the 2017 New Zealand Institute of Architects' National Awards Citation. The project is spatially courageous, reinventing the expected commercial paradigm in favour of a bespoke and spatially fluid approach befitting the unique opportunity of the building. The building addresses public lanes on all frontages, with high levels of operability enabling strong activation of these urban edges. Photos: Dawid Wisniewski and Simon Devitt courtesy of Warren and Mahoney

and a future one projected towards new uses and values that embrace the urban realm and its role in delivering a social sustainability dimension.

Conclusions

This contribution has discussed how, in addition to overcoming decay and preserving the cultural identity of a civilization, the adaptive reuse of existing and heritage buildings has the potential of impacting on the way people perceive and occupy a public space, extending the concept of temporary appropriation towards more complex dynamics where these spaces become a permanent living part of the public realm, populated by ever-evolving purposes. The design project becomes a determining factor in defining the successful outcome of such processes, through the identification of relationships and patterns of behaviours, rather than pre-defined uses. As the architecture of a place is recognised as an enabling factor of temporary appropriation and social interaction, through the definition of values, behaviour and actions (Martínez, 2014), a key to a successful and positive transformation of the built environment is its adaptability to change, understood in a Darwinian meaning of being able to best adapt and adjust to the changing environment and society.

Adaptive reuse has the purpose of indefinitely prolonging the life of a building and tends towards the potential immortality of the building itself and, for this reason, may appear antithetic to temporary appropriation. However, if "the conservation of a place of cultural heritage value is usually facilitated by the place serving a useful purpose" (ICOMOS New Zealand, 2010), then allowing a variety of temporary purposes which, collectively, can retain and reveal those values embedded in the historically significant building could be a successful approach to achieving a development that is socially, economically and environmentally sustainable. In fact, appropriation of space can be limited in time if considering the single activity, but its duration can be easily extended if considering the sum of the actions. It is worth pointing out, though, that this architectural and spatial strategy will need to be assessed in combination with several other criteria, such as retention of both tangible and intangible values, conservation principles, technical feasibility (including code compliance, durability, maintainability, etc.) and, not last, financial feasibility.

The future of architectural heritage depends largely upon its integration into the context of people's lives and upon the weight given to it in regional and town planning and development schemes (Committee of Ministers of the Council of Europe, 1975). In a country with a young building stock as New Zealand, the retention of historically significant landscapes and their integrated purposeful reuse and adaptation has the crucial role of building the heritage of tomorrow, avoiding the loss of cultural values and optimising, preserving or revaluing all the existing social, environmental and economic urban capital.

References

Auckland Council. (2013). Auckland design manual.

Australia ICOMOS. (2013). The Burra Charter. The Australia ICOMOS charter for places of cultural significance.

Boarin, P., Calzolari, M., & Davoli, P. (2016). New interventions in historical and consolidated urban contexts: Low renovation processes for the valorisation of the patina of the time/Nuove dinamiche di intervento nel tessuto urbano consolidato: processi di low renovation per la valorizzazione della. *TECHNE: Journal of Technology for Architecture and Environment, 12*, 103–111. [Online] Available from: doi:http://dx.doi.org/10.13128/Techne-19341.

Brand, S. (1994). *How buildings learn: What happens after they're built.* New York: Viking.

Committee of Ministers of the Council of Europe. (1975). European charter of the architectural heritage.

Conejos, S., Langston, C., & Smith, J. (2011). Improving the implementation of adaptive reuse strategies for historic buildings. In: L. V. dei Mercanti (Ed.), *S.A.V.E. HERITAGE: Safeguard of Architectural, Visual, Environmental Heritage*. (pp. 1–10), Naples.

Douglas, J. (2006). *Building adaptation* (2nd ed.). Oxford, Amsterdam, Boston, London: Butterworth-Heinemann.

English Heritage. (2009). Conservation areas. *Conservation Bulletin*, 62, 1–56.

EU Member States Ministers. (2010). Toledo Declaration.

Filep, C. V., Thompson-Fawcett, M., & Rae, M. (2014). Built narratives. *Journal of Urban Design, 19*(3), 298–316. [Online] Available from: https://doi.org/10.1080/13574809.2014.890043.

Fonseca Rodriguez, J. M. (2015). La importancia y la apropiación de los espacios públicos en las ciudades. *Revista de Tecnologia y Sociedad, 4*(7), 1–11.

Giebeler, G., & Kahlfeldt, P. (2009). Further building work—Thoughts on works with the building stock. In Georg Giebeler, Harald Krause, & Fisch Rainer (Eds.), *Refurbishment manual: Maintenance, conversions, extensions* (pp. 16–19). Walter de Gruyter GmbH: Basel, Berlin, Boston.

ICOMOS. (1964). International charter for the conservation and restoration of monuments and sites (the Venice charter 1964).

ICOMOS New Zealand. (2010). ICOMOS New Zealand charter for the Conservation of places of cultural Heritage Value (p. 11).

Kramer, S. (2015). *Rough style: Architecture, interior, design*. Schweiz, Braun: Salenstein.

Lara-Hernandez, J. A., & Melis, A. (2018). Understanding the temporary appropriation in relationship to social sustainability. *Sustainable Cities and Society, 39C*, 1–14. [Online] Available from: https://doi.org/10.1016/j.scs.2018.03.004.

Lara-Hernandez, J. A., Melis, A., & Caputo, S. (2017). Understanding spatial configuration and temporary appropriation of the street in Latin American cities: The case of Mexico City Centre. In H. Boudagh, A. Versaci, A. Sotoca, F. Trapani, M. Migliore, & N. Clarke, et al. (Eds.), *Advances in Science, Technology & Innovation* (pp. 153–173).

Madanipour, A. (1996). *Design of urban space. An inquiry into socio-spatial process*. New York, USA: Wiley.

Martínez, E. (2014). Configuracion urbana, hábitat y apropiación del espacio. *Scripta Nova. Revista Electrónica de Geografía y Ciencias sociales, 493*(33), 1–20.

Misirlisoy, D., & Günçe, K. (2016). Adaptive reuse strategies for heritage buildings: A holistic approach. Sustainable Cities and Society. [Online] *26*, 91–98. Available from: https://doi.org/10.1016/j.scs.2016.05.017.

New Zealand Productivity Commission. (2015). A history of town planning in New Zealand.

Newman, O. (1973). Defensible space: Crime prevention through urban design. *Ekistics, 36*(216), 325–332.

Plevoets, B., & Sowińska-Heim, J. (2018). Community initiatives as a catalyst for regeneration of heritage sites: Vernacular transformation and its influence on the formal adaptive reuse practice. *Cities, 78*, 128–139. [Online] Available from: https://doi.org/10.1016/j.cities.2018.02.007.

Plevoets, B., Van Cleempoel, K., & WIT Transactions on the Built Environment. (2011). Adaptive reuse as a strategy towards conservation of cultural heritage: A literature review. In: *12th International Conference on Structural Repairs and Maintenance of Heritage Architecture (STREMAH 2011)*, Chianciano Terme (pp. 155–164). [Online]. Available from: https://doi.org/10.2495/str110131.

Purcell, M. (2002). Excavating Lefebvre: The right to the city and its urban politics of the inhabitant. *GeoJournal, 58*, 99–108. [Online] Available from: https://doi.org/10.1023/b:gejo.0000010829.62237.8f.

Relph, E. (1976). *Place and placelessness* (2nd Ed.). London: Pion Limited.

Te Aranga. (2008). Te Aranga Maori Cultural Landscapes Strategy (p. 16).

United Nations Educational Scientific and Cultural Organization. (1972). Convention concerning the protection of the world cultural and natural heritage.

United Nations Educational Scientific and Cultural Organization. (2017). Tangible Cultural Heritage.

Wilkinson, S. J., Remøy, H., & Langston, C. (2014). Sustainable building adaptation: Innovations in decision-making. In H. Therese Remøy, C. Langston, & C. A. Langston (Eds.), Chichester, England: Wiley-Blackwell.

Wong, L. (2017). *Adaptive reuse: Extending the lives of buildings*. Basel Birkhäuser: Basel.

World Commission on Environment and Development. (1987). Our common future: Report of the World Commission on Environment and Development. Annex to UN document A/42/427.

Yung, E. H. K., & Chan, E. H. W. (2012). Implementation challenges to the adaptive reuse of heritage buildings: Towards the goals of sustainable, low carbon cities. *Habitat International, 36*, 352–361. [Online]. Available from: https://doi.org/10.1016/j.habitatint.2011.11.001.

Using the Street in Mexico City Centre: Temporary Appropriation of Public Space Versus Legislation Governing Street Use

Jose Antonio Lara-Hernandez, Alessandro Melis and Claire M. Coulter

Abstract Historically there has been a rich discussion concerning the function of streets in cities, and their role in urban life. This paper outlines the relevance of temporary appropriation for understanding social dynamics within a given urban environment, looking in particular at activities occurring in the street. It takes as a case study Mexico City Centre and examines the laws and regulations set out by the government of Mexico City which regulate the use of the street. It contrasts this with the ways in which the inhabitants of the city appropriate public space on a daily basis. There is a contrast between the lack of clarity in the legislation surrounding potential activities occurring on the street, and a seemingly tacit consensus between citizens regarding how they appropriate such public spaces. We explore this contrast and outline ways in which public space is used in traditional and unexpected ways, how creative ways are found to use the street area within the spirit of the law, and where further research on this topic this could lead in future.

Keywords Mexico City Centre · Public space · City centre · Streets · Temporary appropriation

Introduction

The design of the street as a public space is highly relevant for more than simply physical or aesthetic reasons; it frames our understanding of social encounters in the public realm. The creation of environments that support social behaviour in places is one of the most important roles of urban design (Mehta, 2013).

The need to interact with each other is inborn in humans. According to Mumford (1938), the city is, above all else, a theatre of social action. This suggests the existence of a dialogue between people and their physical environments which is spatialised because activities occur in spaces, and has been given different definitions through time. For instance, street ballet by Jacobs (1961) or the space of contact by Choay (2001), in where all the different social encounters happening in the street, such as

J. A. Lara-Hernandez (✉) · A. Melis · C. M. Coulter
University of Portsmouth, Portsmouth, UK
e-mail: Jose.LaraHernandez@myport.ac.uk

A. Melis et al. (eds.), *Temporary Appropriation in Cities*,
https://doi.org/10.1007/978-3-030-32120-8_13

children playing, people walking, or sitting at a table drinking coffee, mean that public spaces in cities are the physical environment where this interaction between citizens takes place. According to Rywert (Anderson, 1986), the expectation of daily human contact that public space offers is unique. When public spaces do not allow this contact, one of the possible risks is a rise in alienation, which contributes to social stress, unused space, and an increase in crime rates. Moreover, since cars have taken over most of the street space, the only place where the street ballet occurs is on the sidewalk (Minnery, 2012). The street is the immediate public space where urban life is evidenced.

In this chapter we explore how the use of public space is currently regulated in Mexico City Centre (MCC), and how this differs from the ways in which public space has historically been used. We note the discrepancies and loopholes in current legislation, and show how the day to day use of public space in MCC occurs alongside, and sometimes despite, legislation.

With a total population of over 21 million Mexico City forms the core of the fourth largest urban agglomeration in the world, and is both the world's largest Spanish speaking city and the largest city in the Western hemisphere (United Nations, 2014). This is a harbinger of megacity growth in the Global South, where society is profoundly unequal society, wealth is concentrated in few hands and many survive only through informal commerce and construction. People enter the city every day in huge numbers to work or for leisure and other pursuits (Nivon Bolan & Sanchez Bonilla, 2014; Villanueva et al., 2012). Mexico City's sprawl and populace make it a productive choice to explore the use of public space in a megacity, and for showing possible changes in use of public space elsewhere in Latin America. As Herzog (2004) notes, "Mexico City encapsulates what we might call the 'yin/yang' of globalization—it houses both the best and the worst of our global future."

The chapter is divided into three main sections. The first outlines the concept of temporary appropriation (TA), and discusses its relevance as a theoretical concept for understanding the relationship between people and places. The second section attempts to clarify the legal framework that regulates the use of streets in MCC. The third section places the concept of TA within the specific research context, in this case MCC. It explores historical and contemporary uses of public space in MCC, and the varying ways in which TA can be categorised within this research setting. The legal framework for street use stands in contrast to the way in which streets are actually being used, and forms the basis for the discussion of perceived, actual and desired use of public spaces within MCC, viewed through the lens of TA. Finally, we conclude by showing the tension between the popular and legislative use of the streetscape, and suggest areas where this could be further explored in future.

Temporary Appropriation in the Urban Context

Temporary appropriation is relevant as a theoretical concept for understanding the relationship between people and public spaces. Although there is no formal definition of appropriation, other theories incorporate and approach the concept, claiming that it plays a key role in creating the bond between people and places that leads to the social construction of public spaces. In previous work we have explored in depth how TA can offer a valid alternative way of reading the urban landscape (Lara-Hernandez & Melis, 2018), looking at the development of the idea of *appropriation* from its first use in this context in Korosec-Serfaty (1976) and Sansot (1976). We show how more recent work on appropriation, such as the topophilia theory of Yory García (2003, 2011), fits with Lefebvre (1971), who argues that without appropriation, the domination of nature does not have a purpose; there is no urban realm if public spaces are not appropriated. TA is an individual, social, and spatial need that cannot be underestimated when it comes to urban studies, and forms the vital theoretical link between people and places (Lara-Hernandez & Melis, 2018). Fonseca Rodriguez (2015) provides a definition incorporating the temporality of the concept by defining temporary appropriation (TA) as "the act in which people use public spaces to carry out individual or collective activities other than the purpose that the space was originally designed for". This definition helps us to better conceptualise TA as an urban phenomenon.

Looking at the context of MCC, although a considerable amount of literature (Gehl, 2014; Jacobs, 1961; Whyte, 1980) discusses and analyses the physical attributes of the built environment for the purpose of creating lively and inclusive public spaces, studies on the subject in Latin America have only focused on the processes of segregation (Oehmichen, 2010), gentrification (Martinez-Ramirez, 2015), and exclusion (Bayón, 2008) occurring in city centres. Alternative authors, such as Carrion (2013), Garcia Espinosa (2005), Hernández Bonilla and Gómez Gómez (2015) and Hidalgo and Graftieaux (2008) deal with the effects caused by transforming the built environment to create public spaces in city centres in Latin America, however they fail to consider the new public spaces in terms of their ongoing use and TA by the public, let alone exploring the current legal framework that defines the uses permitted for each space in the local context.

Forms of Temporary Appropriation of Public Space

Public spaces are the arenas in which many activities can occur, ranging from leisure pursuits to political protests. Nevertheless, not all activities are the spatialised expression of appropriation. According to Sansot (1976) people only appropriate places which they identify with. Torres (2009) argues that the appropriation of public space is strongly valued by people, since it is culturally constructed by everyday activities. This implies that appropriation of the space plays a key role in people's identities and their interactions with their environment.

Even though a breadth of literature describes the importance of the use of the space in which people, through their activities, appropriate public spaces (Alessandri Carlos, 2014; Fonseca Rodriguez, 2015; Gutierrez de Velasco Romo & Padilla Lozano, 2012; Korosec-Serfaty, 1976), few studies fully specify or classify such activities. Activities relating to trade and commerce commonly occur in public spaces and are associated with appropriation, however there are uncountable other activities occurring besides those linked with profit. Furthermore, while discussing public spaces as a concept we commonly tend to associate this with parks or squares, leaving the street aside. Scholars such as Choay (2001), Gehl (2011), Jacobs (1961), (1993), Kim (2013), Moudon (1991), have described the street as a quintessential public space within the urban realm. The street is a versatile space because even when its meaning is shaped it can be changed, boundaries can be re-defined, new activities can take place and its time management may change too (Mehta, 2013). The space where this conflict is most evident is in public space, commonly in the streets, which is the place that serves as the locus of collective expressions for those who are deprived of institutional settings to disagree (Roy & Alsayyad, 2004). Deleuze and Guattari (1989) suggest that the street is the space for the operation denominated *overcoding* by excellence in which community, state and tradition super impose their code with concrete implications towards the urban realm. The street as urban space is under constant tension between its function as infrastructure and as public space.

As indicated previously there is an infinite number of activities that public spaces could host, however we attempted limit our investigation to those activities in the street which can be classified as those where people appropriate public space. Only by identifying and classifying these forms of TA will it be possible to achieve a deeper understanding of such activities, and their relationship with the street as a public realm. The broad categories that we identified are explored in detail in the paragraphs which follow.

Economic Actions and Temporary Appropriation

According to Ramirez Kuri (2010), there are three types of TA related to commerce or services. First, people such as vendors or workers use the space with or without the authorisation of government authorities; second, people work in public spaces as an extension of their place of work because it is close by; lastly, people gather in the public space, with clients who travel there to find them and use their services, such as mariachi bands. Gutierrez de Velasco Romo and Padilla (2012) agree with the identification of these three types of the occurrence of TA of public spaces in central areas. Even though informal commerce on the street is usually perceived as an undesirable activity by governments in different cities (Kim, 2013; Ramirez-Lovering, 2008), it is without any doubt an activity in which people worldwide temporarily appropriate the street.

Leisure and Temporary Appropriation

Activities related to leisure, such as arts or sport, are also evidence of TA. Cranz (1982) claimed that recreational and leisure activities occurring in public spaces, such as parks, have a strong influence on urban political processes in North American culture. Crouch (1998) explores the significance of the street as an everyday site of geographical knowledge and leisure practice, revealing the rituals and relationships, and practices and representations which are played out routinely on the street. He also states that the design of both the streets and the buildings dictate the experience of a place. Both authors imply that through leisure and recreational activities, people appropriate public spaces; it is emotionally beneficial when people participate in leisure activities in public spaces because they feel comfortable to do so, but it is even more important that they feel as if they own the space.

Sacralisation and Temporary Appropriation

In countries with a strong religious cultural background, another activity in which people appropriate public spaces is through sacralisation. Portal (2009a) describes the term sacralisation by referring to religious symbols (mostly Catholic) being placed in a public space. This form of appropriation is particularly characterised by the installation of crosses or altars in public spaces, where people may pray. According to Portal (2009a), there are two main causes for this kind of appropriation. First, as an act of personal or familial commemoration, for instance because a friend or relative has died on the site or close to it. Portal (2009a) notes that for violent deaths, a cross can be seen as a way of helping the soul to find peace on its onward journey. In other situations, flowers and other artefacts can be placed on the site to mark the place where a death occurred, appropriating the space for personal commemoration. The second cause to mark or define territories at the boundary of a determinate area or neighbourhood, and can be sued to reduce anti-social behaviour within a designated area (Portal, 2009a, 2009b). Santino (1992) explores this idea further, noting that artefacts forming public displays of commemoration not only invite spectatorship, but through their invitation to bear witness, involve passers-by otherwise unconnected to the event in the act of TA, as outlined by Habermas (1991) through his ideas on shared civic interest.

Summary

People make use of public space for a variety of reasons, and to undertake a wide assortment of activities. We have aimed to categorise these as simply as possible under the broad headings of economy, leisure and sacralisation, but appreciate that under each of these headings a complex network of activities is taking place. Table 1 provides an overview of the activities in which people temporarily appropriate public spaces. In the section which follows we will explore how these forms of activity take place within MCC, looking at specific examples from historical sources as well as from the present.

Table 1 Activities in the public space related to temporary appropriation (*Source* Lara-Hernandez, Melis, & Caputo, 2017)

Category	Economy		Leisure			Sacralisation
Description	Any activity in which a person or group use the public space in order to obtain an economic benefit directly or indirectly		Any activity in which a person or a group use the public space for leisure purposes			Any activity in which a person or a group use the public space for religious purposes
Sub-category	Work	Trade	Sports-games	Artistic expressions	Rest	
Individual or collective	Advertising or promoting services, waiting, engaging or attracting possible clients	Selling or buying products (food, handicraft, clothes, etc.)	Skateboarding, soccer, cards, marbles, hopscotch	Playing music, dancing, painting, acrobatics, reciting and singing	Eating, resting	Processions, praying, lighting candles and putting flowers

The Official Use of the Street in Mexico City Centre

We analysed the laws and regulations approved by the government of Mexico City regarding the use of the street using document analysis (Bowen, 2009). In addition, in order to gather further information about how the street (as public space) is actually managed, an interview was conducted with the general coordinator of *Autoridad del Centro Historico*, aimed at collecting information on temporary appropriation and whether it is being considered at street management level.

Mexican Laws and Regulations Regarding Street Use

Several laws and regulations in Mexico City (Asamblea Legislativa del Distrito Federal III Legislatura, 2004; Gobierno de la Ciudad de México, 2014; Gobierno del Distrito Federal, 2013, 2015, 2016) refer to the use of streets and the activities that are allowed to happen in them. Table 2 summarises the information found in Mexican laws and regulations.

In 2004, the *Ley de la Cultura Civica del Distrito Federal* (Law of the Civic Culture of Mexico City) was approved by the government, and establishes how citizens should behave in public spaces and neighbourhoods. The 15th article of Chapter I/Second Title states that it will guarantee the harmony and coexistence of its inhabitants through the fulfilment of their duties, such as (a) the freedom of people's actions in public spaces, and (b) by allowing the proper use of public spaces according to their nature.

In 2013 a more specific law was approved: the *Ley para el uso de las vías y los espacios públicos del Distrito Federal* (Law for the use of streets and public spaces in Mexico City), which establishes the right to use and enjoy public spaces, especially streets that are used in different ways, rather than streets being just for motor-vehicles. Article 6 states that users of public space (including streets) have the right to use the property for common use according to its nature, and have the right to access, stay and transit in streets. Article 7 establishes that public space users have the duty to access, use, stay in, or transit through public space without disturbing other users. Both laws clearly refer to the social dimension of streets as public spaces.

Other laws, such as *Reglamento de Transito del Distrito Federal* (Transit Regulations of Federal District) and *Ley de Movilidad de la Ciudad de Mexico* (Law of Mobility of Mexico City) establish the free access, use and transit of pedestrians, cyclists and motor-vehicles in streets and roads. They also establish that streets should be free of obstacles or elements that impede or hinder pedestrian traffic, except in authorised cases. Both laws give priority to pedestrian movement; and they acknowledge different uses for the street beyond just movement.

Moving beyond laws aimed at traffic regulation and mobility, socio-urban researchers, such as Ramirez Kuri (2016), Campos Cortés and Brenna Becerril (2015) and Ramirez Kuri (2010), argue that the codes compiled by planners and

Table 2 Laws and Regulations about the use of streets in Mexico City (translations: Authors)

Name	Law for the use of streets and public spaces in Mexico City (Gobierno del Distrito Federal, 2013)	Law of the Civic Culture of Mexico City (Asamblea Legislativa del Distrito Federal III Legislatura, 2004)	Law of Mobility of Mexico City (Gobierno del Distrito Federal, 2016)	Transit Regulations of Federal District (Gobierno del Distrito Federal, 2015)	Citizen Manual for the care of City Centre (Gobierno de la Ciudad de México, 2014)
Generalities and functions	Stabilised the right to use and enjoy the public space, including streets used in a different way rather than for motor-vehicles. The public space is considered as an ambience or scenery of social integration, where the right of association and the right of the others to use the same space, its appropriation (accessibility, permanence and enjoyment) of the space, the collective space, the space for everybody	It establishes the minimum rules of citizen behaviour	It dictates the laws for mobility purposes in Mexico City	It regulates the use of streets and roads of Mexico City	It is a manual for citizens that use MCC

(continued)

Table 2 (continued)

Name	Law for the use of streets and public spaces in Mexico City (Gobierno del Distrito Federal, 2013)	Law of the Civic Culture of Mexico City (Asamblea Legislativa del Distrito Federal III Legislatura, 2004)	Law of Mobility of Mexico City (Gobierno del Distrito Federal, 2016)	Transit Regulations of Federal District (Gobierno del Distrito Federal, 2015)	Citizen Manual for the care of City Centre (Gobierno de la Ciudad de México, 2014)
Specific articles that mention the use of public space	Art 6th. The users of public space and streets have the right to I. Use the property of common use according to its nature and destiny: II. To access public spaces and streets: III. To stay in public spaces and streets; IV. To transit in public spaces and streets	Second Title of the Civic Culture and neighbour's participation. Chapter I. Art 15. The Civic Culture in Mexico City, which guarantees the harmonious coexistence of its inhabitants, is based on the fulfilment of the following citizen duties. VI. To allow the freedom of action of people in public spaces and streets. X. Make proper use of public goods, spaces and services according to their nature and destination	Art 5. Mobility is the right of every person and the community to carry out the effective movement of individuals and goods to access through the different modes of transportation recognized in the Law, a mobility system that conforms to the hierarchy and principles that are established in this order, to meet their needs and full development. In any case, the object of the mobility will be the person	Art. 1. The purpose of these regulations is to regulate the circulation of pedestrians and vehicles on public roads and road safety in Mexico City. Art. 6.- The Public Administration will provide the necessary means for people to freely choose the way to move in order to access the goods, services and opportunities offered by the City. The level of vulnerability of users, the externalities generated by each mode of transport and their contribution to productivity will be considered for the establishment of public policy in this area. Priority will be given to the use of road space and the distribution of budgetary resources will be assessed according to the following hierarchy of mobility: I. Pedestrians	Chapter 7. Use and conservation of public spaces. Conditions for realising cultural activities. First, permission must be granted by the SSPDF (Secretary of Public Safety of Mexico City), the SGDF (Mexico City Council) and Autoridad del Centro Historico

(continued)

Table 2 (continued)

Name	Law for the use of streets and public spaces in Mexico City (Gobierno del Distrito Federal, 2013)	Law of the Civic Culture of Mexico City (Asamblea Legislativa del Distrito Federal III Legislatura, 2004)	Law of Mobility of Mexico City (Gobierno del Distrito Federal, 2016)	Transit Regulations of Federal District (Gobierno del Distrito Federal, 2015)	Citizen Manual for the care of City Centre (Gobierno de la Ciudad de México, 2014)
	Art 7. The duties of the users are: I. To use the public space and streets according to their nature and destination: II. To exercise their rights without disturbing the order and public peace, or affecting the developing continuity of other inhabitants' activities		Art 13. For the fulfilment of the present Law and the ordinances that emanate from it, Public Security will have the following attributions: III. Maintain within the scope of its powers, that the road is free of obstacles and elements that impede, hinder or impede vehicular and pedestrian traffic, except in only authorized cases, in which case, as far as possible, they should not obstruct the accesses for persons with disabilities		

urban designers have focused on the economic development and formal aspects of the area, rather than including lessons from other fields dealing with social and cultural aspects. Urban design interventions have been implemented according to a master plan, titled *Plan Integral de Manejo del Centro Histórico de la Ciudad de México* (Management Plan for México City Centre), which involves pedestrianisation, change of pavement, sidewalk expansion, and the addition of urban furniture as well as lighting and trees. According to Flores Arias (2015), even though physical improvements of public spaces in MCC are the result of a plural approach (involving the opinion of the academic sector, government, experts and general population), they are implemented in a top-down manner, and as a result, do not represent the interests of the population (2015). Though aesthetically pleasing for tourists and visitors, the urban design interventions conducted in MCC in conjunction with changes to the urban landscape (physical and social) have resulted in the eviction of urban actors, thus eliminating the lively social dynamics that they were contributing to Campos Cortés and Brenna Becerril (2015), Martinez-Ramirez (2015). These planning policies are just as significant for the use of public space as the laws cited in Table 2, and yet they are far less accessible and therefore less open to debate and to discussion.

The View of the Authorities on Temporary Appropriation

As stated above, we conducted an interview with a key figure in the local city administration to collect a fuller picture of the official understanding of TA in the streetscape. The answers provided in Table 3 help to give a picture of the government's working policy on TA activities. On the one hand, it clearly shows that there is an aversion towards economic activities related to work, and particularly towards trade and commerce. On the other hand, activities related to leisure such as sports and games, artistic expression, resting, religious activities and pedestrian use are strongly welcomed. The single most striking observation to emerge from the data was that many of the activities which were reported as desirable could actually be categorised as obstacles for pedestrians, and consequently convene the official legal position. As a result there is some discrepancy between the written legal and the day-to-day official views regarding activities that are welcome to occur in the street in MCC.

Summary

Mexico City law says public spaces must be accessible by all citizens with no distinction or impediment (Gobierno del Distrito Federal, 2013). Mexican citizens may legally appropriate any public space or street, and transport is not the only reason streets are there. Citizens must also respect other users of the spaces and their right to use the area for their own purposes, so long as pedestrian traffic is not impeded and the peace is not disturbed. Commercial activities are not mentioned or banned

Table 3 Questionnaire response from Autoridad del Centro Historico[a]

To what extent is it desirable for these activities to happen on the street?					
	Very desired	Moderately desired	Somewhat	Minimally desired	Not at all
Activities					
Work	5	4	3	2	1
Trade	5	4	3	2	1
Sport/games	5	4	3	2	1
Artistic expressions	5	4	3	2	1
Rest	5	4	3	2	1
Religious activities	5	4	3	2	1
Pedestrian use	5	4	3	2	1
Demographic study					
Do you live in MCC?				(y)	(n)
Do you work in MCC?				(y)	(n)
Do you live in San Jerónimo St?				(y)	(n)
Do you work in San Jerónimo St?				(y)	(n)
Do you live in Moneda St?				(y)	(n)
Do you work in Moneda St?				(y)	(n)

[a]The Faculty of CCI Ethics Committee reviewed the questionnaire (Appendix 4) prior the study

and no restrictions are placed on any type of activity that can be peacefully enjoyed apart from an inconclusive reference to the use of streets being according to their nature.

This is not how the authorities would prefer streets to be used. Official policy would rather restrict commercial activities and see transit and leisure as a street's primary purposes. No priority was placed on maintaining pedestrian access and some preferred activities would in fact impede it. Official urban planning policies give tourist experiences preferential treatment and frequently disregard local residents' and habitual street users' needs.

Appropriation of Streets in Mexico City Centre

Mexico City Centre: Historical Use

Mexico City Centre (MCC) is the biggest colonial historic centre in America, with a total area of 9.1 km^2 and a population of 6122 (INEGI, 2018). In terms of planning, MCC has the Spanish grid commonly used in Latin American colonies, starting from the main square, the church and the council building (see Fig. 1). The streets of Mexican cities are public spaces that are key for urban life, even prior to the Spanish colonisation. The pre-Hispanic civilisations were characterised by the intensive utilisation of the outdoor spaces for everyday activities (Keller, 2006; Suárez Pareyón, 2004). The Aztecs founded the city of Tenochtitlan, the urban pattern of which was compounded by blocks, streets and channels. The Aztecs used the channels for communication purposes, while streets were reserved for a diversity of activities such as trade, leisure, religious celebrations and even sacrifices (Leon Portilla, 1995). The public spaces in which the social, political, economic and religious lives of people

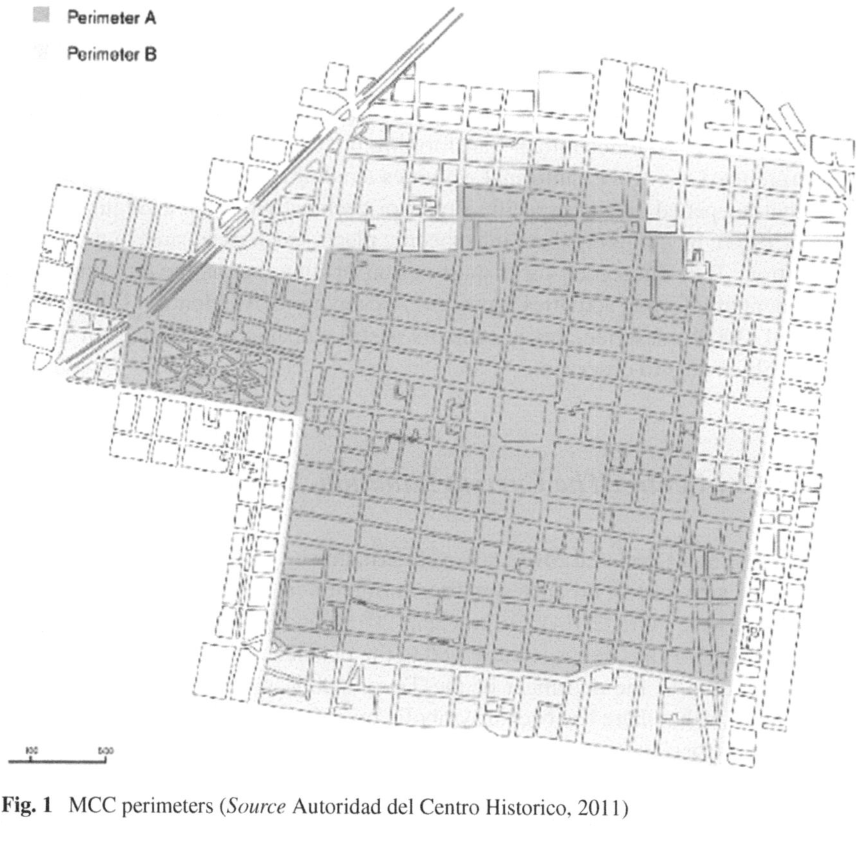

Fig. 1 MCC perimeters (*Source* Autoridad del Centro Historico, 2011)

Fig. 2 The great Tenochtitlan, mural by Diego Rivera Exhibited in the Palacio Nacional in Mexico City (*Source* Rivera, 1945)

occurred were the streets (Webster & Sanders, 2001). Informal activities were an essential element of street life in Tenochtitlan, especially for the common people, as shown in Fig. 2 (Crossa, 2009).

In the 16th century, during the Spanish colonisation, a new urban planning pattern was established over Tenochtitlan, transforming the city (Stanislawski, 1947). The conquerors tried to regulate informal activities (trading, playing, religious expressions) that happened on each specific street for each activity with a singular order (Nelson, 1963), by confining them into specific places, such as squares in the Spanish tradition. They succeeded for a short period of time, but as the city grew, the confinement of informal activities was not viable anymore. As Monnet (1995, 1996) and Portal (2009b) describe, this informality is still palpable in the streets of MCC, and is easily visible to observers.

Mexico City Centre: The Current Situation

In 1987, UNESCO declared MCC a world heritage site. Although the declaration has triggered a process of urban design transformation intended to beautify the built environment, Díaz Parra (2014), Oehmichen (2010), and Ramirez Kuri (2015) have pointed out that this has not necessarily led to the planned improvements urban life, and has occurred to the detriment of the TA of public spaces. Hiernaux-Nicolas (2005) suggests that there has been a symbolic privatisation of heritage in favour of the dominant class and commercial recovery, challenging the strategies promoted by the entrepreneurial urban governance and in doing so creating counter-spaces. As an example of this, one of the strategies of the transformation program was the removal of street vendors from perimeter A (see map in Fig. 1) (Autoridad del Centro Historico, 2011). As a result of this action, the use of public spaces in MCC has changed, with a collateral effect on the relationship between urban design and TA of public spaces. As part of our research into the use of public space (Lara-Hernandez & Melis, 2018; Lara-Hernandez, Melis, & Caputo, 2017) photographs were taken in MCC during 2017 showing different activities taking place in the street and in public squares. Many of these activities clearly show public space being temporarily appropriated by members of the public, and fall into the three categories discussed previously. We also found instances of official use of public space which fell under TA, as well as activities which were unexpected, and which defied our original categorisation. The three main categories, as well as the outliers, will be discussed in turn below.

Economic Actions and Temporary Appropriation

Looking at economic forms of temporary appropriation, there is evidence of disruption or disarrangement to the everyday practices of residents, indigenous artisans, and many local vendors, impacting directly on the economy of many families. This has altered the traditional Mexican way of living in cities, which even before the Spanish colonisation was and is about socialising and trading outdoors. According to Crossa (2009), even though the strategies implemented by MCC's *Programa de Rescate* have attempted to remove informal commerce from streets, street vendors have found ways to resist, and they have become *toreros* (a term referring to Mexico City's nomadic vendors), and they are still working in the area. For instance, street vendors in Moneda St. place their merchandise on a cloth on the ground that can easily be gathered up if necessary (see Fig. 3 left). A "guardian" with a walkie-talkie is constantly looking out for the presence of authorities (see Fig. 3 right). Once the "guardian" has spotted the authorities he notifies the street vendors who immediately

Fig. 3 Street vendors in Moneda St. (left), Street vendors' guardian with walkie-talkie (right) (*Source* Authors)

Fig. 4 Street vendors hiding in buildings nearby (left), street vendors have left the street (right) (*Source* Authors)

collect their merchandise and go and hide in a building nearby (see Fig. 4 left). In a matter of seconds the whole temporary market has vanished (see Fig. 4 right).

As we have seen, the law regarding street use in MCC primarily protects the rights of pedestrians to have access to the street for transit. In further examples of commercial driven TA, citizens as both consumers and entrepreneurs have found ways to work around the law, for example by placing small chairs and tables right next to the edge of a building (see Fig. 5 left). The street which remains a viable place for transit thus simultaneously also becomes an open-air dining area. Similarly, in the example illustrated in Fig. 5 (right), a hair treatment is being carried out using a bollard as a hairdresser's chair, with the street becoming briefly an urban salon. Moreover, even employees of governmental institutions such as the police or refuse collectors temporarily appropriate the street for their own purposes. Figure 6 (left) shows police officers holding their morning meeting in the street, while Fig. 6 (right) shows a line of refuse collectors lining up to receive their weekly payment. In both of these examples people temporarily appropriate the street, using the space as an urban open office.

Fig. 5 People eating on the street (left), woman doing hair treatment on the street (right) both in Moneda St. (*Source* Authors)

Fig. 6 Police officers in an ordinary morning meeting (left), refuse collectors getting their weekly payment (right) both in San Jeronimo St. (*Source* Authors)

Leisure and Temporary Appropriation

Looking at leisure activities, Delgadillo (2014) describes the planning policies that have been implemented as part of the urban beautification process. *La Alameda Central* (the largest park in MCC), used to be appropriated by families, minority groups, religious groups, mimes and even musicians. After the urban design transformation that took place in 2013, this type of appropriation is not happening anymore (Martinez-Ramirez, 2015). The new policies allow the government to displace users who are perceived to be informal, suspicious or misbehaving; to pursue an official cultural programme of free cultural and leisure events which prevent the free use and public dimension of this space; and to follow a zero-tolerance agenda, including an increase in police officers and surveillance in the area (Delgadillo, 2014).

In contrast to the traditional, family-centric and group-oriented use of the street, many of the examples of leisure related TA that we observed in MCC were more individual in nature. In the examples shown in Fig. 7 we see a man napping and a

Fig. 7 Man in wheelchair taking a siesta (left), Woman knitting (right) (*Source* Authors)

Fig. 8 Open-air workshop (left) woman playing the violin (right) (*Source* Authors)

woman knitting; two individuals make use of a small peripheral area at the edge of a street or square, and their chosen activity does not invite interaction with the wider public. Figure 8 shows more complex use of the street area, and highlights how TA can take multiple forms, in this case both economic and leisure-based. In the left hand picture a workshop takes place outside a church, in the right hand picture a woman is playing the violin in the street. Although both of these activities could plausibly be primarily economic in nature, their ability to be enjoyed or participated in by the wider public extends them beyond merely commercial forms of TA.

Sacralisation and Temporary Appropriation

The third category of TA that we expected to find in MCC relates to religious activities. Whilst we observed traditional street altars occupying space within the street (see Fig. 9), we also found unexpected instances of TA which could be categorised under sacralisation. The celebration of *Día de los Muertos* (day of the dead) in Mexico is a well-known as a family gathering to remember deceased family and friends. It is a national celebration in which the whole community participates. Although prostitution is a practice which is not commonly associated with family values, the prostitutes of *La Merced* (a neighbourhood adjacent to MCC) gather together on this day and make their own public *ofrenda* (Castrejon-Arcos, 2012; Redacción ADN, 2015). Figure 10 illustrates how despite the laws, regulations and the social stigma, the prostitutes of *La Merced* temporarily appropriate the street to celebrate *Día de los Muertos.*

Fig. 9 Street altar in Mexico City Centre (right) (*Source* Authors)

Fig. 10 Prostitutes celebrating *Dia de los Muertos* in La Merced (*Source* Castrejon-Arcos, 2012)

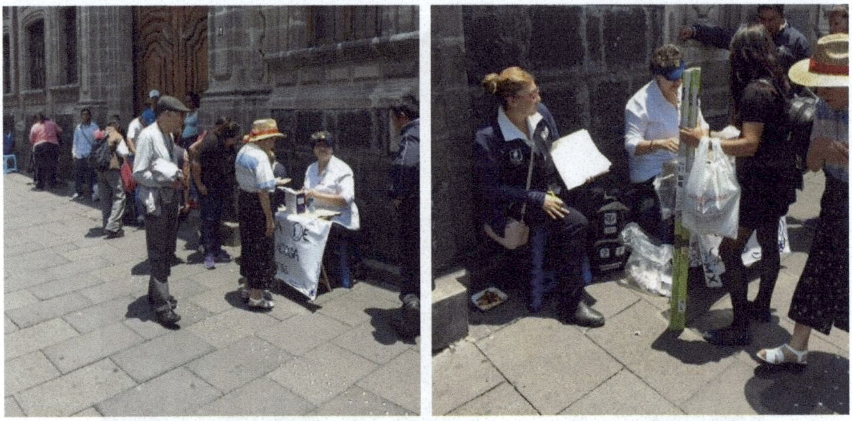

Fig. 11 Nurses conducting tests to diagnose diabetes (*Source* Authors)

Other Categories of Temporary Appropriation

Although our previous work had suggested three definitive categories for TA, we found other examples of street use in MCC that did not fit neatly into any of these categories. Firstly, we observed that some health institutions temporarily appropriate the street, providing their services in the open-air for the benefit of the citizens (see Fig. 11).

Another example are the daily protests that take place in Mexico City (see Fig. 12). Between 2015 and 2017 there were more than ten thousand protest events registered by *Secretaria de Seguridad Publica* (City Security Department) and *Secretaria de Gobierno* (City Council), an average of nine protests every single day (Arredondo, 2018). By law (Gobierno del Distrito Federal, 2013) citizens willing to take part in a protest utilising the public spaces (squares or streets) must to give notice to the authorities 72 h prior the event. The authorities however state that just over half of these events receive advance notification, highlighting how the citizens' use of public spaces is embedded as a right in the context of Mexico City.

Discussion

Temporary Appropriation of the Street, and the Laws and Regulations of MCC

As mentioned in the conceptual discussion, certain activities occurring in public spaces are considered TA, playing a key role in creating the bond between people and places, and consisting broadly of activities relating to commerce, leisure and

Fig. 12 Protest against the education bill (*Source* Authors)

religion. The current economic situation in MCC, whereby the built environment has been maintained and improved, may have had a positive impact on the tourist experience, but has arguably had a less favourable impact on everyday socio-urban conditions, which according to Van der Aa (2005), is a condition of many other heritage sites worldwide.

Some authors (Alessandri Carlos, 2014; Ramirez Kuri, 2008; Saraví, 2008) claim that planning regulations imposed on the use of public space are putting the inclusive and pluralistic nature of these spaces at risk. This situation poses a threat to the TA of public spaces, which according to Purcell (2002), is one of the key ways for citizens to exercise their right to the city. Inclusion, diversity and governance are fundamental urban issues that will not be resolved by space where everyone is not treated as equal. Legislation governing street use in MCC is loose, only reserving rights of pedestrian transit and keeping the peace. There is tacit recognition in laws and regulations of the right to temporarily appropriate public space. Only when we examine internal policy and communication does discussion begin to surface about what forms of TA are desirable and appropriate. The city's plans for urban regeneration are centred on leisure activities as the area's key activity driver, with as much concern for maintenance of the historic centre for visitors as for the enjoyment of the local population. Furthermore, our interview with director of a governmental institution showed that activities related to leisure and religion were considered desirable, whilst commercial activity was to be discouraged, even though the demographic section of the questionnaire shows that the respondent is fully aware of the urban dynamic happening in the area. In contrast to the arguments made by Delgadillo (2014) Janoschka and Sequera (2014) and Ramirez Kuri (2015, 2016) that regulations have been imposed on public spaces that hinder TA we have instead found that the regulations themselves are vague and lack definition, and that it is internal planning policies, that come under far less public scrutiny, which are forming the backbone of the effort to modify street use in the area.

The Changing Use of the Street

Although the use of public space is changing, it is significant that the urban actors have not been totally evicted, rather the way in which they temporarily appropriate the streets has changed. Whilst we observed the expected categories of economic, leisure and religious activities, we saw this appear in unexpected ways, as well as ways which we were unable to fit into these categories. We observed how TA is not only occurring in the acts of private citizens, but also in a semi-official context. Police meetings, governmental staff pay queues, and publicly administered healthcare were all observed happening in the street. Likewise we were surprised to observe acts of TA from groups that would otherwise be considered controversial, but which were permitted within a specific time or context. The example of the prostitutes celebrating on *Dia de los Muertos* highlights just how flexible the boundaries of TA can be, whilst the use of the street area for public protests reinforces what we found in the legislation, that citizens have the right to appropriate public space as long as they do so peacefully and preserve pedestrian access.

The legislation that we surveyed was not particularly specific on how the public may make use of the street, although internal governmental policies on street use were much clearer. Similarly the finding that around half of public protests do not notify the authorities in advance, but apparently without any issues, suggests that where TA is concerned there is a form of legal pluralism occurring in MCC, where state law, religious law, indigenous law, customary laws and local conventions all co-exist, and combine to govern the actual use of the street. In such plural realm the law of the state is not necessarily the dominant one. Furthermore, the state might not actually have the capacity to enforce the law (McAuslan, 2006). In this scenario customary law could have much more influence on how activities are conducted in public, with the written law, conventionally taken as formal, being considered informal and beneath consideration in practice.

Our research's key finding is that, while some activities could be considered illegal, nothing happened that failed to respect the law's spirit. Pedestrian traffic was respected by informal commerce, prostitutes did not disturb the peace when celebrating a national holiday, there was sharing of political views and there were governmental meetings and no-one was seen to question the right to use the street in any of these ways. While some activities were seen as more desirable than others, no official consensus exists on limits to permitted TA in MCC, and the streets are used for a wide range of activities.

Conclusion

This chapter has shown that TA of the street is crucial for maintaining socio-urban dynamics in the contemporary city. The literature on TA was used in an attempt to divide TA activities in MCC between economic actions, leisure activities and religious practices, but activities on the streets were much wider than that and also included healthcare, political activity and official government business. In comparison with a legal position that does not mention permitted activities, and regeneration policies that saw informal commercial activity as undesirable, in day-to-day use the street reflects the local population's practical needs and wants, however creative they need to be in finding ways to conduct their business.

Some authors have noted a decrease in certain types of TA within MCC, leading to a change in the way that the street is used and a loss of inclusivity. The official laws and regulations appear not to be the cause of this, but there is some conflict between how the authorities address the management of public space and how people have previously used these areas. It is in any case unclear whether changes to the built environment in MCC can be considered improvements and have helped to maintain TA while preserving the site's social heritage, or whether it has created new life on the streets to the benefit of a small sector of society, or even only for tourists and other outsiders. The role of time in changing the use of public spaces and their TA is essential, and requires further research.

A limitation of our research is that it only focusses on the laws and regulations for the use of the streets (and public spaces) in a particular area in the centre of Mexico City. It does not address fully how the streets are actually used and managed as public spaces within the city as a whole. Further research into the city's planning strategies, policies on urban street use, and the actual use of the street in other districts is needed to confirm our findings that the use of space is affected both by state law and local conventions, changes to which could have an impact on citizens' social obligations (Lara-Hernandez, Melis, & Lehmann, 2019).

Future studies on the topic in other cities and multicultural environments are also recommended. Cities are in a constant process of transformation, and the way in which they are designed and re-designed might affect the rational use of public spaces and their TA. A total of 68% of the world population will be living in urban areas by 2050 (United Nations, 2018), with urban areas seeing a corresponding increase in cultural diversity alongside population growth. Van Hook (2017) estimates that in the USA the diversity has increased by 98% in urban areas since 1980. These socio-urban changes represent a challenge for urban design and planning. For instance, Pemberton (2016) points out that there is a lack of work about the role of urban planning towards diversity and migration. We have seen that TA has cultural components which may be specific to a specific place or ethnic group. Research is needed to investigate how cities that are more culturally diverse and/or which are facing high levels of national and international migration (such as London, Barcelona or Sydney) address the issue of the use of public space. It would be interesting to see how the forms of TA practiced in each city vary according to the presence of different cultural and religious groups living in the same area, and how this either coexists or comes into conflict with local legislation governing the use of the street.

References

Alessandri Carlos, A. F. (2014) La ciudad como privación y la reapropiación de lo urbano como ejercicio de la ciudadanía. *Scripta Nova. Revista electrónica de geografía y ciencias sociales* (Vol. XVIII, No. 493, pp. 1–13). ISSN 1138-9788.

Anderson, S. (1986). On streets. In *On streets*. Cambridge, Massachusetts, UK: MIT Press.

Arredondo, Í. (2018). *9 marchasal día por 3 años*. El Universal. Retrieved February 1–5 from http://www.eluniversal.com.mx/periodismo-de-investigacion/cdmx-9-marchasal-dia-por-3-anos.

Asamblea Legislativa del Distrito Federal III Legislatura. (2004). *Ley de Cultura Cívica del Distrito Federal. Gobierno Del Distrito Federal* (pp. 1–29). Retrieved from http://www.aldf.gob.mx/archivo-0f05874fac7a0a4b94b9935dd0998eae.pdf.

Autoridad del Centro Historico. (2011). *Plan Integral de Manejo del Centro Historico de la Ciudad de Mexico (2011–2016)*. Mexico, DF.

Bayón, M. C. (2008). Desigualdad y procesos de exclusión social. Concentración socioespacial de desventajas en el Gran Buenos Aires y la Ciudad de México. *Estudios Demográficos y Urbanos, 23*(1(67)), 123–150. https://doi.org/10.2307/40315269.

Bowen, G. A. (2009). Document analysis as a qualitative research method. *Qualitative Research Journal, 9*(2), 27–40. https://doi.org/10.3316/qrj0902027.

Campos Cortés, G. I., & Brenna Becerril, J. E. (2015). Repensando el espacio público como un bien común urbano. *Argumentos. 28*(77), 157–176. http://www.redalyc.org/articulo.oa?id=59540679008.

Carrion, M. F. (2013). Dime quién financia el centro histórico y te diré qué centro histórico es. *Revista Digital del Programa en Gestion de la Ciudad., 16,* 1–26.

Castrejon-Arcos, D. E. (2012). *Sexo servidoras de la Merced. Ofrenda de día de muertos.* DECA-Logos. http://deca-logos.blogspot.com/2012/10/ofrenda-de-dia-de-muertos-de-las-sexo.html. Accessed 17 January 2019.

Choay, F. (2001) *The invention of the historic monument.* Cambridge, UK: Cambridge University Press. https://doi.org/10.2307/991857.

Cranz, G. (1982) *The politics of park design: A history of urban parks in America.* Cambridge, UK: MIT Press.

Crossa, V. (2009) Resisting the entrepreneurial city: Street vendors' struggle in Mexico City's historic center. *International Journal of Urban and Regional Research, 33*(1), 43–63. https://doi.org/10.1111/j.1468-2427.2008.00823.x.

Crouch, D. (1998). The street in the making of popular geographical knowledge. In *Images of the street: Planning, identity and control in public space* (pp. 160–175).

de Gutierrez Velasco Romo, E. G., & Padilla Lozano, F. (2012). Apropiación y uso del espacio público en el centro de la ciudad de Aguascalientes. *Investigación y Ciencia., 20*(56), 33–43.

Deleuze, G., & Guattari, F. (1989). In B. Massumi (Ed.), *A Thousand plateaus: capitalism and schizophrenia.* Minneapolis, MN: University of Minnesota Press. https://doi.org/10.2307/203963.

Delgadillo, V. (2014). La Política del espacio público y del patrimonio urbano en la Ciudad de México. Discurso progresista, negocios inmobiliarios y buen comportamiento social. In *XIII Coloquio Internacional de Geocrítica.* Barcelona, España: Universitat de Barcelona.

Díaz Parra, R. I. (2014). La transformación del centro histórico de la Ciudad de México. ¿Gentrificación o repoblación? In *XII Coloquio y Trabajos de Campo del Grupo de Geografía Urbana (AGE). Territorios inconclusos y sociedades rotas* (pp. 1–10). 2014 Madrid y Castilla la Mancha.
Flores Arias, V. (2015). El Centro Histórico de la Ciudad de México y sus retos en el siglo XXI. In *Planur-e*. http://www.planur-e.es/articulos/ver/el-centro-hist-rico-de-la-ciudad-de-m-xico-y-sus-retos-en-el-siglo-xxi/completo. Accessed 1 May 2016.
Fonseca Rodriguez, J. M. (2015). La importancia y la apropiación de los espacios públicos en las ciudades. *Revista de Tecnologia y Sociedad., 4*(7), 1–11.
García Espinosa, S. (2005). Centros históricos ¿herencia del pasado o construcción del presente? Agentes detonadores de un nuevo esquema de ciudad. Scripta Nova. *Revista electrónica de geografía y ciencias sociales, IX*(39), 1–11.
Gehl, J. (2011). *Life between buildings*. Washington, DC: Island Press.
Gehl, J. (2014). *Places for people*. Melbourne, Australia: City of Melbourne.
Gobierno de la Ciudad de México. (2014). Manual ciudadano para el cuidado del centro histórico. In *Fideicomiso Centro Histórico de la Ciudad de México* (p. 95). http://www.autoridadcentrohistorico.cdmx.gob.mx/storage/app/uploads/public/58a/490/da3/58a490da3b8ca502567977.pdf.
Gobierno del Distrito Federal. (2013). *Ley para el uso de las vías y los espacios públicos del Distrito Federal*. Diario Oficial de la Federación. https://doi.org/10.1017/cbo9781107415324.004.
Gobierno del Distrito Federal. (2015). *Reglamento de Transito del Distrito Federal* (p. 124). http://www2.df.gob.mx/virtual/tlahuac/transparencia/fraccion1/REGLAMENTOS/REGLAMENTODETRANSITODELDF.pdf.
Gobierno del Distrito Federal. (2016). *Ley de Movilidad del Distrito Federal* (pp. 1–101). http://www.paot.org.mx/centro/leyes/df/pdf/2016/LEY_MOVILIDAD_DISTRITO_FEDERAL_07_10_2016.pdf.
Habermas, J. (1991). *The structural transformation of the public sphere: An inquiry into a category of Bourgeois Society*. Mass, MI: MIT Press. https://doi.org/10.2307/2072652.
Hernández Bonilla, M., & Gómez Gómez, S. (2015). Conflicto espacial, exclusión y espacio público en la centralidad urbana de Xalapa, Veracruz. *Revista Brasileira de Gestão Urbana. 7*(3), 281–294. https://doi.org/10.1590/2175-3369.007.003.ao04.
Herzog, L. A. (2004) Globalization and the crisis of public space: The example of Mexico. In: G. A. Jones & P. M. Ward (Eds.), *The end of public space in the Latin American City*? (pp. 5–8). Austin, TX: Mexican Center of LLILAS. http://lanic.utexas.edu/project/etext/llilas/cpa/mexcenter/endofpubspace.pdf.
Hidalgo, D., & Graftieaux, P. (2008). Bus rapid transit systems in Latin America and Asia: Results and difficulties in 11 Cities. *Transportation Research Record: Journal of the Transportation Research Board. 2072*(2072), 77–88. https://doi.org/10.3141/2072-09.
Hiernaux-Nicolas, D. (2005). *Transformaciones morfológicas de la Ciudad de México* (pp. 9–22). Revista Veredas.
INEGI. (2018). Clima. Distrito Federal. Cuentame... http://cuentame.inegi.org.mx/monografias/informacion/df/territorio/clima.aspx?tema=me&e=09. Accessed 30 July 2018.
Jacobs, J. (1961). The death and life of Great American. *Cities* : https://doi.org/10.2307/794509.
Jacobs, A. B. (1993). *Great Streets: Monument Avenue, Richmond, Virginia* (pp. 22–28)
Janoschka, M., & Sequera, J. (2014). Procesos de gentrificación y desplazamiento en América Latina - una perspectiva comparativista. Desafíos metropolitanos. *Un diálogo entre Europa y América Latina., 2014,* 82–104.
Keller, A.H. (2006). A road by any other name: trails, paths, and roads in Maya language and thought. In *Landscapes of Movement Symposium* (p. 29).
Kim, A. M. (2013). *Sidewalk city*. Chicago, IL: The University of Chicago Press. https://doi.org/10.7208/chicago/9780226119366.001.0001.
Korosec-Serfaty, P. (1976). Appropriation of space. In *Proceedings of the Strasbourg Conference, IAPC-3*. Strasbourg-Lovaine La Neuve: CIACO.

Lara-Hernandez, J. A., & Melis, A. (2018). Understanding the temporary appropriation in relationship to social sustainability. *Sustainable Cities and Society, 39C*, 1–14. https://doi.org/10.1016/j.scs.2018.03.004.
Lara-Hernandez, J. A., Melis, A., & Caputo, S. (2017). Understanding spatial configuration and temporary appropriation of the Street in Latin American cities: the case of Mexico City Centre. In H. Boudagh, A. Versaci, F. Trapani, & M. Migliore, et al. (Eds.), *Advances in science, technology & innovation* (pp. 153–173).
Lara-Hernandez, J. A., Melis, A., & Lehmann, S. (2019). Temporary appropriation of public space as an emergence assemblage for the future urban landscape: The case of Mexico City. *Future Cities and Environment, 5*(1), 1–22. https://doi.org/10.5334/fce.53.
Lefebvre, H. (1971). De lo rural a lo urbano. In J. Gonzalez-Pueyo (Ed.), Barcelona, EspañaL: Ediciones Peninsula.
Leon Portilla, M. (1995). *Antologia de Tehotihuacan a los Aztecas. Fuentes e interpretaciones históricas. First re-i.* Mexico, DF: Universidad Nacional Autonoma de Mexico.
Martinez-Ramirez, U. (2015). Proceso de Gentrificación y Dezplazamiento en el espacio público del Centro Histórico de la Ciudad de México. WPCC-15006.
McAuslan, P. (2006). Legal pluralism as a policy option: Is it desirable, is it doable? *Land Rights for African Development From Knowledge to Action, 2005*, 9–10. http://www.capri.cgiar.org/wp/..%5Cpdf%5Cbrief_land-04.pdf.
Mehta, V. (2013). *The street. A quintessential social public space*. New York, NY: Routledge.
Minnery, J. (2012). Sidewalks: Conflict and negotiation over public space. *Urban Policy and Research. 1146*, 1–2. https://doi.org/10.1080/08111146.2012.729694.
Monnet, J. (1995). *Usos e imágenes del centro histórico de la Ciudad de México*. Centro de estudios mexicanos y centroamericanos (ed.).
Monnet, J. (1996). Espacio público, comercio y urbanidad en Francia, México y Estados Unidos. *ALTERIDADES, 6*(11), 11–25.
Moudon, A. V. (1991). *Public streets for public use*. Columbia University Press: New York, NY.
Mumford, L. (1938). The culture of cities. In *The culture of cities* (pp. 405–481). London, UK: Anthony Rowe.
Nelson, H. J. (1963). Townscapes of Mexico: An example of the regional variation of townscapes. *Economic Geography, 39*(1), 74–83. http://www.jstor.org/stable/142496.
Nivon Bolan, E., & Sanchez Bonilla, D. (2014). La gestión del Centro Histórico de la Ciudad de Mexico: 1980–2012. In L. Durán, E. K. Garcés, & M. Lacarrieu (Eds.), *Nuevos aportes al debate desde America Latina* (pp. 48–65). Quito, Ecuador: Durán, Lucía.
Oehmichen, C. (2010). Espacio urbano y segregación étnica en la ciudad de México. *Cuadernos de estudios sociales urbanos. 6*, 61–75. http://www.cedib.org/wp-content/uploads/2013/05/villalibre_6.pdf#page=62.
Pemberton, S. (2016). *The challenge of superdiversity for urban planning*. UPWEB. https://www.birmingham.ac.uk/generic/upweb/blog/2016/the-challenge-of-superdiversity-for-urban-planning.aspx. Accessed 18 January 2019.
Portal, M. A. (2009a). Las creencias en el asfalto. La sacralización como una forma de apropiación del espacio público en la ciudad de México. *Cuadernos de Antropología Social., 30,* 59–75.
Portal, M. A. (2009b). Memoria y territorio: la sacralización simbólica del espacio urbano. In *La Historia Oral y la multidisciplinariedad: retos y perspectivas* (pp. 1–9). Colima, Mexico: Departamento de Antropología UAM-1.
Purcell, M. (2002). Excavating Lefebvre: The right to the city and its urban politics of the inhabitant. *GeoJournal, 58*, 99–108. https://doi.org/10.1023/b:gejo.0000010829.62237.8f.
Ramirez Kuri, P. (2008). La fragilidad del espacio publico en la ciudad segregada. In R. Cordera, P. Ramirez Kuri, & A. Ziccardi (Eds.), *Pobreza, desigualdad y exclusión social en la ciudad del siglo* (Vol. XXI, pp. 1–25). Mexico, DF: Siglo XXI editores.
Ramirez Kuri, P. (2010). La ciudad y los nuevos procesos urbanos. *Sociología Urbana,* 163–187.
Ramirez Kuri, P. (2015). Espacio público, ¿espacio de todos? Reflexiones desde la ciudad de México. *Revista mexicana de sociologia., 1*(77), 7–36.

Ramirez Kuri, P. (Ed.). (2016). *La reinvencion del espacio publico en la ciudad fragmentada*. Mexico City: Universidad Nacional Autonoma de Mexico.
Ramirez-Lovering, D. (2008). *Opportunistic urbanism*. Melbourne, Australia: RMIT Press.
Redacción ADN. (2015). Inversión por 16 MDP en parques públicos y espacios deportivos en Oaxaca: Administración (20:40 h) ‹ADN – Agencia Digital de Noticias Sureste. ADN Sureste. http://adnsureste.info/inversion-por-16-mdp-en-parques-publicos-y-espacios-deportivos-en-oaxaca-administracion-2040-h/. Accessed 28 January 2019.
Rivera, D. (1945). *La Gran Tenochtitlan*. Commons Wikimedia. https://commons.wikimedia.org/wiki/File:La_Gran_Tenochtitlan.JPG [Accessed: 29 August 2017].
Roy, A., & Alsayyad, N. (2004). *Urban informality: Transnational perspectives from Midle East, Latin America and South Asia*. Oxford, UK: Lexington Books.
Sansot, P. (1976). Notes sur le concept d'appropriation. In *Conférence Internationale de Psychologie de l'Espace Construit (3rd IAPS) de Strasbourg* (pp. 67–75). Strasbourg, France.
Santino, J. (1992). Yellow Ribbons and Seasonal Fla The Folk Assemblage of War. *The Journal of American Folklore., 105*(415), 19–33.
Saraví, G. A. (2008). Mundos Aislados: Segregación Urbana y Desigualdad en la Ciudad de México. *EURE (Santiago), 34*(103), 93–110. https://doi.org/10.4067/s0250-71612008000300005.
Stanislawski, D. (1947). Early Spanish Town planning in the new world. *American Geographical Society, 37*(1), 94–105.
Suárez Pareyón, A. (2004). *El centro histórico de la Ciudad de México al inicio del siglo XXI. Revista invi., 19*(51), 75–95.
Torres, E. (2009). *Apropiación Versus Usos Del Espacio Público*. https://www.mardelplata.gob.ar/documentos/policia/taller1culturayterritorioapuntes.pdf. Accessed 1 November 2017.
United Nations. (2014). World's population increasingly urban with more than half living in urban areas [Online]. Department of Economic and Social Affairs. Available from: http://www.un.org/en/development/desa/news/population/world-urbanization-prospects-2014.html. Accessed 29 September 2015.
United Nations (2018). *68% of the world population projected to live in urban areas by 2050, says UN*. United Nations Department of Economic and Social Affairs (UN DESA). https://www.un.org/development/desa/en/news/population/2018-revision-of-world-urbanization-prospects.html. Accessed 17 September 2018.
Van der Aa, B. J. M. (2005). *Preserving the heritage of humanity? Obtaining world heritage status and the impacts of listing*. Groningen, The Netherlands: Rijksuniversiteit.
Van Hook, J. (2017) Diversity is on the rise in urban and rural communities, and it's here to stay. In *The Conversation*. https://theconversation.com/diversity-is-on-the-rise-in-urban-and-rural-communities-and-its-here-to-stay-69095. Accessed 18 January 2019.
Villanueva, N., Vazquez, R., Hernandez, S., Chiu, J., et al. (2012). Metodología para la Construcción del Sistema de Indicadores de Prevención Social de la Violencia y la Delincuencia Septiembre de 2012, Mexico, DF.
Webster, D., & Sanders, W. (2001). La antigua ciudad mesoamericana: teoría y concepto. *Urbanismo En Las Sociedades Antiguas*, 43–64. http://dialnet.unirioja.es/servlet/articulo?codigo=2776058.
Whyte, W. H. (1980). *The social life of small urban spaces*. https://doi.org/10.1177/089124168201000411.
Yory, M. C. (2011). El Concepto de Topofilia entendido como Teoría del Lugar. *Revista Barrio Taller*, 1–17.
Yory García, C. M. (2003). *Topofilia, ciudad, territorio: una estrategia pedagógica de desarrollo urbano participtivo con dimensión sustentable para las grandes metrópolis de América Latina en el contexto de la globalización: 'el caso de la ciudad de Bogotá'*. Universidad Complutense de Madrid. http://biblioteca.ucm.es/tesis/ghi/ucm-t26725.pdf.

Transforming Everyday Public Space: Human Appropriations in Search for Citizenship and Urban Well-Being

Luisa Bravo

Abstract A new urbanity pervades the public life of our cities, based on the search for "ordinary magic" in everyday social life. Spontaneous actions, practices and interventions promoted by various groups, mostly private, are able to redesign the spaces of collective life, through the use of creative ideas and tools. Following a do-it-yourself urbanism approach, working on the potential of places that are often banal, empty areas not completely designed or underused spaces that are not immediately perceptible for collective use, contemporary urban makers or performers are able to create new places, linked to the multiple needs of community life. The results, often temporary, have the ability to impose themselves strongly on the community's imagination, from an emotional and participatory point of view, because they can demonstrate what the city could be, while spreading a collective feeling of urban well-being.

Keywords Do-it-yourself urbanism · Co-city · Human appropriations · Urbanity

Introduction

In 2005 the architectural firm Rebar in San Francisco transformed a single metered parking space into a temporary mini public park in the SOMA district, that was known to be underserved by public open space. Just by rolling out living grass, putting up a bench and placing a potted tree, they created an opportunity for social interaction for two hours, inviting passers-by to sit and relax. As Rebar described 'within minutes, a man sat down on the bench, took off his shoes, and began to eat lunch, another person joined soon after, and the two began having a conversation'. What was a space for cars became a temporary place for people. Rebar then packed up and returned the space to its former condition.

At that time, 70% of San Francisco's downtown outdoor space was dedicated to the private vehicle, while only a fraction of that space was allocated to the public realm. But after the Rebar's installation, the idea to reuse the parking space for

L. Bravo (✉)
City Space Architecture, Bologna, Italy
e-mail: luisa.bravo@cityspacearchitecture.org

A. Melis et al. (eds.), *Temporary Appropriation in Cities*,
https://doi.org/10.1007/978-3-030-32120-8_14

public use soon became the expression of an alternative thinking, reclaiming public space as a fundamental value for urban life. After that, Rebar prepared an how-to manual, addressing the technical aspects of site specific installations, intended to share what they called 'an act of generosity' with anyone was interested to create his/her own park. Later Rebar received support from the Trust for Public Land and the temporary appropriation and transformation of a parking spot went global, and was celebrated as Park(ing) Day on a set day of the year. The actors of this transformation were artists, designers and citizens, committed to affirming the need for greater urban space for public life: in extremely creative ways and forms, they covered the asphalt with green carpets and set up tables and chairs in cafés and restaurants at open air, in a well-defined space separated from vehicular traffic, through low-cost design interventions. The event was so successful that it was replicated in many cities around the world, in independent forms from the original project but based on shared guidelines: in 2011 the Park(ing) day involved 975 locations, 162 cities, 35 countries, 6 continents, inspiring new ways to improve public space. In 2008 the City of San Francisco approached Rebar to prototype 'parklets', as more permanent Park(ing)-style installations, sponsored by local businesses, and later to start the Pavement to Parks program (Bravo et al., 2018). That's how temporary appropriations became permanent outdoor public spaces.

Urbanity on Demand: DIY Urbanism

Increasingly, people and groups around the world are bringing incremental improvements to streets, blocks and neighbourhoods through small-in-scale informal urban design and spontaneous interventions of micro-urbanism. Design activists are private groups and organizations determined to advocate a better city and society, they work independently or in collaboration with property owners, developers, planners and politicians. In response to a considerable reduced influence of local governments in terms of spatial production and large scale projects, due mainly to a significant financial crisis, these small-in-scale, bottom-up and temporary initiatives have been rapidly elevated to mainstream practice. With a view to acting for the well-being of the community, with limited economic resources but engaging full creativity and cooperation, they invest their time and their ability with disinterested generosity, characterized, almost always, also by a playful component that invites participation, without age distinctions, and becomes the engine of consensus, participation and approval. They utilise existing spaces or require minimal investment, infusing places with value and meaning. Often the results are temporary but they can have a great impact on residential communities and are able to foster imagination and care.

At the 13. Venice Architecture Biennale in 2012, the United States of America pavilion's exhibition '*Spontaneous Interventions. Design Actions for the Common Good*' showed 124 projects representing this urban activism and received a special mention for national participation from the Biennale's Golden Lion jury (spontaneousinterventions.org). Those projects celebrated widespread desires for good places and for the freedom to improve everyday public life, even when there is no client, no architect and no authority to sanction them (Lang Ho, 2012), so that they could be even claimed as illegal. They are related to a new democratic design movement looking for the "ordinary magic" of the so called "everyday space", namely the physical domain of everyday public activity, made of ordinary places that function as "connective tissue binding daily lives together" (Chase & Crawford, 2008: p. 6). These places can be ambiguous, obvious, banal, un-designed or invisible, such as vacant lots, sidewalks, front yards, parks, and parking lots (Ben-Joseph, 2012) serving as public space for private, commercial and domestic purposes, but they have the potential to foster new forms of social interaction, even if they work as collective places only a few hours during the day or only a few days during the week or the month. These spaces, from suburbia to the inner city, constitute an "Everyday Urbanism", continually shaped by communities and redefined by their transitory activities, rather than by a proper design intervention.

Urban transformations are more and more related to aspirations of citizens and based on negotiation practices and sensory experiences, developed through temporary and ever-changing opportunities for public life. As Bunschoten points out, "the public realm is the playground in which society reinvents itself" (2002: p. 6): temporary urbanism, also called guerrilla, do-it-yourself, bottom-up or pop-up urbanism, is redefining the concept of public space through the establishment of "open-minded spaces". In opposition to single-minded spaces, designed and used for a specific purpose, open-minded spaces can acquire constantly changing meanings, social, aesthetic, political, economic, and are designed for a variety of uses, including unforeseen and unforeseeable ones, for citizens who do different things and are prepared to tolerate, even take an interest in, things they don't do (Walzer, 1986).

These "open-minded spaces" reverse conventional concepts of "public" and "space": instead of reproducing the morphology of "public spaces", they refer to the activities of different "publics", observing the temporal rhythms and daily itineraries that define their spaces. The concept of "public", as defined by sociology and political sciences, namely "the public", refers mainly to populations of individuals, sharing a common interest, involved in civic affairs. In architecture and urban design, the idea of "public", often used in the plural/multiple form, "publics", is related to new forms of social engagement in the public realm related to different actors or groups, and are often linked to "commons", intended as public goods and shared resources in the urban context. The relation between actors and places, namely between "publics" and "commons", has become particularly intriguing in the last decades, when it started to shape a new kind of urbanity, based on aspirations of equity and citizenship through participation, made with creative contents.

The social life of different "publics" is able to produce places, namely "urban commons", with community significance (Bravo & Crawford, 2014), fostering social engagement and spreading a collective feeling of urban well-being. Imagination and the creative ability to re-think the urban environment and the public domain and to live it in unconventional and unusual ways (Hou, 2010) are the key ingredients for a successful revitalization of the urban space. As Jaime Lerner, the well-known former Mayor of Curitiba in Brazil, pointed out, 'the lack of resources is no longer an excuse not to act'. While major transformations usually require a significant top-down effort, in terms of engagement of experts, analysis and studies, planning and design, implementation and construction, these easy-to-do and long-lasting temporary and bottom-up initiatives are unexpectedly lighter, quicker and cheaper. Lerner's dictum that 'creativity starts when you cut a zero from your budget' has widely inspired the hearts and minds of urban and social entrepreneurs all around the world.

Temporary Occupations Need Space: Approaching the Human Scale

Temporary urbanism is making clear that public space is not merely a design activity, rather it is the opportunity to unlock aspirations of ownership and citizenship of different publics. Public life is more and more the product of occasional urbanities (Peran, 2008) derived by unpredictable forms of human engagements through small-in-scale practices. Urban agriculture, temporary stores, art performances, daily and seasonal community events, mobile trucks for urban food, street vending, kiosks, sport and recreational activities, re-use of vacant shops and derelict building sites are becoming more and more frequent activities in many public spaces all around the world, promoted and managed by local communities (Beekmans & de Boer, 2014) Local governments are starting to invest in public space, to provide opportunities for these activities to flourish, unleashing the potential embedded in different places, activating and regenerating both social and urban contexts and generating cultural values.

The start-up 'Urb-I Urban Ideas' (https://www.urb-i.com/) based in Brazil has collected, starting from 2015, more than 3000 pictures of urban transformation of places worldwide, from car-oriented to a more pedestrian friendly design, in a before/after gallery generated using the time machine of Google street view. This gallery clearly shows how an under-designed urban space still embeds a great potential for public life, that could be activated by redefining priorities, putting pedestrians first, and by providing opportunities for temporary urbanism (Madanipour, 2017). The founders of Urb-I are 'working towards the democratization of urbanism', as they state on their website, 'in the creation of better places for all people'. In the US, the so-called tactical urbanism promoted by The Street Plans Collaborative (Lydon & Garcia, 2015), namely a citizen-led approach for low cost and short term transformation, has quickly developed into inspiring and scalable activities able to catalyse long-term change (tacticalurbanismguide.com), through many open-source documents and books, including a 'How to' guide, aimed at inspiring and empowering self-organized individuals and communities.

In Copenhagen the architectural firm Gehl, initiated in 2000 by Jan Gehl (1987) with Helle Søholt, in the past 18 years worked globally to implement a pioneering vision, through research, projects and consultancy: making cities for people. This human scale approach is a radical change to the current urban mind-set that is hard to take root. For the first time in 2010, the number of vehicles registered worldwide has exceeded 1 billion units, in the United States of America 239.8 million units constitute the largest volume of vehicles circulating in the world. According to the World Health Organization, between 2010 and 2013 there was a 16% increase in the number of registered vehicles in the world. While the car industry is reinventing a different market, with electric and autonomous cars, a new generation of urban leaders, designers and activists are pushing local governments to adopt human scale policies, reclaiming the urban space for public use, shifting from a car-oriented to a human-centred model. Richard Sennett, the Centennial Professor of Sociology at the London School of Economics, advocates unresolved narratives, porosity, incompleteness and flexibility, so that social richness and complexity can always find a place to flourish in the urban spaces of the city. He argues that the top-down city belongs to the masters, the open city is a bottom-up place and belongs to the people.

At the beginning of 2014, Theatrum Mundi, a charitable incorporated organisation based in UK, in partnership with the American Institute of Architects New York, launched "Designing for Free Speech" challenge: they asked architects, designers, activists, artists - and anyone interested in imagining new spaces in the city for free expression - to identify public spaces in the city of New York and propose re-designs that transform them into places that activate the rights enshrined in the First Amendment of the United States Constitution. Applicants were required to propose architectural or performative designs, temporary or permanent, that transform spaces into places for public demonstration. This challenge was about re-imagining and idealizing existing spaces with the potential to animate the public, especially spaces that are not traditionally considered in this frame (designingforfreespeech.org).

Even in a situation of crisis, where decision-makers experience the limits of economies and ecosystems, individuals and communities can be proactive in generating a change, taking responsibilities for their cities and engaging in improving their environment. The London School of Economics (LSE Cities) and the Alfred Herrhausen Society (AHS) since 2007 developed a five-year research, as an outcome of the Urban Age conference series, on 'Handmade Urbanism' in five cities in emerging countries, Mumbai, São Paulo, Istanbul, Mexico City, Cape Town, as a part of the Deutsche Bank Urban Age Award. This research showed how community-driven actions can be carried out at the local scale through participatory initiatives, as products of a specific culture and context, dealing with soft infrastructure, such as physical and emotional wellbeing or education, and with the reshaping and taking ownership of the built environment (Rosa & Weiland, 2013).

In 2014 at the MOMA Museum in New York, the exhibition "Uneven Growth: Tactical Urbanisms for Expanding Megacities" brought together six interdisciplinary teams of researchers and practitioners to work on new architectural possibilities for six global metropolises: Hong Kong, Istanbul, Lagos, Mumbai, New York, and Rio de Janeiro (https://uneven-growth.moma.org). The exhibition was seeking "to challenge current assumptions about the relationships between formal and informal, bottom-up and top-down urban development, and to address potential changes in the roles architects and urban designers might assume vis-à-vis the increasing inequality of current urban development" and was intended to analysed "how emergent forms of tactical urbanism can respond to alterations in the nature of public space, housing, mobility, spatial justice, environmental conditions, and other major issues in near-future urban contexts" (Gadanho et al., 2014).

Co-designing Cities: Engaging the Experts

The 2016 International Architecture Biennale Rotterdam (IABR) on 'The New Economy' included many projects showing how new coalitions of individuals and collectives are progressively stepping in for a conspicuously absent government. The public domain is at the heart of this revolution, since it can ensure an 'inclusive, resilient and agile' urban future, according to the curatorial statement. Incorporating temporary and bottom-up initiatives into planning and top-down decision making would require a drastic change: metropolitan systems should support this new economy while allowing for production of small-scale manufacturing, social enterprises, creative and informal practices, blending permanent with temporary. This would require a new and different expertise and the ability to define an area of interaction in that in-between space where top-down and bottom-up meet, in order to create a common ground of shared knowledge, able to inform the decision-making process. As Fred Kent, the founder of Project for Public Spaces based in New York and the initiator of the Placemaking movement, would say, ordinary people are the real urban experts, "they know what they like and what they don't; where they will spend time and where they won't, and increasingly they know how to get what they want" (Michaels, 2011).

A group of American and Italian scholars have recently conceptualized the 'Ciy as a Commons' (Foster & Iaione, 2016), based on extensive investigation and experimentation of new forms of collaborative city-making, as part of the LabGov, Laboratory of the governance of the city as a commons (labgov.city). The project defines new frontiers of participatory urban governance, inclusive economic growth and social innovation, through peer-to-peer production, participatory design and cooperation among the actors of the quintuple helix model, i.e. social innovators, public authorities, businesses, civil society organizations, and knowledge institutions. The project, rooted in the conceptual pillars of the urban commons, have studied and collected several ground-breaking experiments, such as FabCity transition plan towards re-localized and distributed manufacturing of Barcelona; the Bologna Regulation on Public-Civic Collaboration for the Urban Commons; the Turin Co-City policy; San Francisco, Seoul and Milan initiatives to transform themselves into "sharing cities"; and Edinburgh as a "cooperative city", and surveyed 67 cities globally over 18 months, in order to build a Co-City index and measure the implementation of the EU and UN Urban Agenda (commoning.city). So the Co-city model is working on the enhancement of social welfare and on the implementation of the right to the city through governance. As Harvey pointed out, 'the freedom to make and remake our cities and ourselves is one of the most precious yet most neglected of our human rights' (Harvey, 2008: p. 23).

The right to the city is far more than the individual liberty to access urban resources: it is a common right to change ourselves by changing the city.

Conclusion: Defining a New Urban Paradigm

The system of rules and major interventions for urban development and land use is nowadays undermined by new agile and fresh bottom-up initiatives promoted by active and self-organized citizens: animated by a willing determination to give a contribution in the process of city-making, with the capacity to act in response to a specific question in a specific context, relying on reduced but well-managed economic resources, these urban design activists are revamping civic pride and the right to the city, through temporary occupations and creative transformations of the ordinary, so called "everyday space", namely the network of empty or residual areas not completely designed or underused spaces that are not immediately perceptible for collective use. Architects, policy makers and city managers, the official experts, are called to confront themselves with the unpredictable dynamics produced by these bold, new community experts, who are passionately reclaiming public space as a common good, giving new meanings to the everyday space through social engagement and creative innovation.

Temporary and grassroots appropriations, based on creative and participatory practices, are filling the gaps left by inefficient planning policies or by the poor design of proper spaces and services. In the Global South, where large populations live in slum-like conditions, such activities can have an enormous impact on urban residents, fostering democratic engagement and social inclusion, nurturing and activating the local human potential. In the Global North, local governments are starting to support these projects, through the definition of a specific framework of policies and activities, where, instead of just taking advantage of unpaid and committed volunteers, they could define a collaborative environment aimed at co-designing spatial interventions and supporting spontaneous, bottom-up initiatives.

Temporary urbanism is defining a new urban paradigm, based on openness, multiplicity, imagination and collaboration. The transformation of public space is defined: through a multi-scalar and multidimensional approach, from a comprehensive governmental city-wide strategy to citizen-led tactical interventions at the human scale; through the combination of different knowledge, engaging conventional experts with community-based experts and urban thinkers; through an innovative process of participatory practices, for the co-production of solutions able to respond to the real, complex needs of urban residents.

References

Beekmans, J., & de Boer, J. (2014). *Pop-up city. City-making in a fluid world*. Amsterdam: BIS Publishers.

Ben-Joseph, E. (2012). *Rethinking a lot. The Design and Culture of Parking*. Cambridge: MIT Press.

Bravo, L. (2018). The urban lounge and a green view. *The Journal of Public Space*, *3*(2), 173–178. https://doi.org/10.32891/jps.v3i2.1115.

Bravo, L. & Crawford, M. (2014). Publics and their spaces: Renewing urbanity in city and suburb. In *New urban configurations* (pp. 784–789) [Online]. October, 2014. Available from: https://doi.org/10.3233/978-1-61499-365-0-789.

Bunschoten, R. (2002). *Public spaces*. London: Black Dog Publishing.

Chase, J. L., Crawford, M., & KaliskI, J. (Eds.). (2008). *Everyday urbanism*. New York: Monacelli Press.

Foster, S. R., & Iaione, C. (2016). The city as a commons. *Yale Law & Policy Review, 34*(2), 281–349 [Online]. Available from: https://digitalcommons.law.yale.edu/ylpr/vol34/iss2/2.

Gadanho, P., Burdett, R., Cruz, T., Harvey, D., et al. (2014). *Uneven growth: Tactical urbanisms for expanding megacities*. New York: The Museum of Modern Art.

Gehl, J. (1987) *Life between buildings. Using public space*. New York: Van Nostrand Reinhold (first edition in Danish, 1971).

Harvey, D. (2008) The right to the city. *New Left Review, 53*(September–October), 23–40.

Hou, J. (2010). *Insurgent public space: Guerrilla urbanism and the remaking of contemporary cities*. New York: Routledge.

Lang Ho, C. (2012). *Spontaneous interventions: Design actions for the common good,* in Architect (pp. 20–26). The Magazine of the American Institute of Architects [Online]. Available from: https://www.architectmagazine.com/design/spontaneous-interventions-design-actions-for-the-common-good_o.

Lydon, M., & Garcia, A. (2015). *Tactical Urbanism. Short-term action for long-termchange*. Washington: Island Press.

Madanipour, A. (2017). *Cities in time. Temporary urbanism and the future of the city*. London: Bloomsbury.

Michaels, S. (2011). *A conversation with Fred Kent, leader in revitalizing city spaces*. The Atlantic [Online]. Available from: https://www.theatlantic.com/national/archive/2011/09/a-conversation-with-fred-kent-leader-in-revitalizing-city-spaces/245178/. Accessed: July 15, 2019.

Peran, M. (2008). *Post-it city. Occasional urbanities*. Barcelona: CCCB.

Rosa, M., & Weiland, U. E. (2013). *Handmade urbanism: From community initiatives to participatory models: Mumbai, São Paulo, Istanbul, Mexico City, Cape Town*. Berlin: Jovis.

Walzer, M. (1986). Public space-pleasures and costs of urbanity. *Dissent, 33*(4), 470–475.

General Conclusion

Jose Antonio Lara-Hernandez, Alessandro Melis and James Thompson

Abstract This book sets out to explain and describe TA as an urban phenomenon which is a socio-spatial need, it has shown TA as a crucial link between people and places, having a multi-scalar nature deeply intertwined with other dimensions of the urban landscape such as environmental, socio-cultural, economic and political. It has identified three main groups of activities (but not limited to) in which TA could be categorised such as the ones related to economy (work or trade), leisure (sports and arts) and sacralisation. In terms of assemblage theory, TA emerges as a product of other assemblages such as culture, legal framework and urban design of public spaces. The relevance of this relationship between these three assemblages is noticeable when looking at the TA diversity index as an indicator of inclusiveness. The element of time within TA is essential to understand the constitution of collective subjectivity such as informal practices but is usually underestimated by urbanists and planners. In terms of crime prevention, it has been discussed that urban design practices following the CPTED principle of "Activity Support", seems to have a negative impact on overall TA while a positive impact on crime reduction in the area of intervention. This tacit identification process through TA leads to a socially sustainable use of the public space, stimulating local economies, supporting household livelihoods, enhancing civic engagement and promoting urban identity. Moreover, TA promotes the right to metabolism, either individually and communally, is essential to achieve a conciliation between urban spaces and nature-society relations in the benefit of people. It explains how TA could help to evidence formerly hidden social practices, meanings, and flaws of ongoing development patterns, and thus be leveraged and operationalised to inform the design of community engagement practices and built environment solutions, such informal settlements. The insights gained from the studies presented here may be of assistance to architects, urban designers and planners because the spatial configuration of a place is recognised as a triggering factor of TA and social interaction, through the definition of values, behaviour and actions.

J. A. Lara-Hernandez (✉) · A. Melis
University of Portsmouth, Portsmouth, UK
e-mail: Jose.LaraHernandez@myport.ac.uk

J. Thompson
University of Melbourne, Melbourne, Australia

A. Melis et al. (eds.), *Temporary Appropriation in Cities*,
https://doi.org/10.1007/978-3-030-32120-8_15

Keywords Temporary appropriation · Public space · Urban design

This book sets out to explain and describe TA as an urban phenomenon which is a socio-spatial need, having a multi-scalar nature deeply intertwined with other dimensions of the urban landscape such as environmental, socio-cultural, economic and political. Chapter "Understanding Temporary Appropriation and Social Sustainability" has identified TA as a crucial link between people and places, understood as "the temporary act in which people use public spaces to carry out individual or collective activities other than the purpose that the space was originally designed for" (Fonseca Rodriguez, 2015). It has identified three main groups of activities (but not limited to) in which TA could be categorised such as the ones related to economy (work or trade), leisure (sports and arts) and sacralisation. In addition, Ta has been suggested as an indicator of the social sustainability of the urban landscape. As an attempt to construct a theory of TA, Chapter "Between Assemblages and Temporary Appropriation: The Case of Mexico City" has conceptualised TA in terms of assemblage theory, illustrating that TA emerges as a product of other assemblages such as culture, legal framework and urban design of public spaces. The relevance of this relationship between these three assemblages is noticeable when looking at the TA diversity index as an indicator of inclusiveness. The highest diversity index of TA was observed in areas where the design of the built environment was neither the best quality or the worst, it was somehow in between, highlighting the crucial role of the urban design for creating more inclusive public spaces.

Thus, Chapters "Understanding Temporary Appropriation and Social Sustainability" and "Between Assemblages and Temporary Appropriation: The Case of Mexico City" have provided the conceptual framework, pushing forward TA as a research topic in its own right despite being commonly misunderstood as a type of informality occurring in public spaces, criticising the narrow understanding of informality within the academic arena as explained in Chapter "Temporary Appropriation and Informality". In general, therefore it seems that informality is a multidimensional concept across time and space, which is complex enough to allow development of case specific research. Moreover, the element of time within TA is essential to understand the constitution of collective subjectivity such as informal practices but is usually underestimated by urbanists and planners, already discussed by Keith Harris in Chapter "Ongoing Appropriation: Invisible Seattle and Red May".

Urban design practice have direct repercussions on TA of public spaces which has been shown by Alexander Gomez-Torres in Chapter "Temporary Appropriation and Public Space: Assessing the CPTED Principle of Activity Support". It has been discussed that urban design practices following the CPTED principle of "Activity Support", seems to have a negative impact on overall TA while a positive impact on crime reduction in the area of intervention, which is the Third Millennium Park in Bogota DC, Capital District of Colombia. Thus, such design principles seems to expel crime from the intervened site rather than promote TA, contrasting common assumptions by urban design practice, aiming to increase and promote peoples activities as a tool to prevent the occurrence of crime (Gehl, 2011; Jacobs, 1961;

Minnery, 2012). Furthermore, this crime reduction could be interpreted as crime displacement, because of the increment of crime incidence throughout the site's neighbouring areas. The study lays the groundwork for future research into TA and its potential contribution to crime reduction.

TA is also influenced by climate and comfort temperatures. In fact, environmental conditions seems to be one of the primary factors that enable TA of public spaces as Daniele Santucci, Ata Chokhachian and Thomas Auer demonstrated in Chapter "Temporary Appropriation of Public Spaces: The Influence of Outdoor Comfort". People experience the built environment in a dual dimension psychologically and physiologically, both strictly related to TA. In other words, TA occurs in public spaces where people feel culturally and environmentally identified. This tacit identification process through TA leads to a socially sustainable use of the public space, stimulating local economies, supporting household livelihoods, enhancing civic engagement and promoting urban identity, such as the example of labourers such as *jornaleros* discussed by Sean Crotty in Chapter "Origins of Informality: Examining the Historical and Spatial Roots of Informal Day-Labor Hiring Sites". It evidenced the multi-scalar nature of TA, which on the one hand showing the socioeconomic linkages at local level while on the other illustrating the migratory movements at global level, leading to the emergence of TA. Furthermore, some expressions of TA are product of interaction, ensuring that access to space is maintained, fostering relationships that produce socially-sustainable outcomes.

Turning now to the other spectrum of the socioeconomic occupation of the urban landscape, in Chapter "Unsheltered Homelessness and the Right to Metabolism: An Urban Political Ecology of Health and Sustainability" Jeff Rose illustrates TA of public spaces by unsheltered homeless, that is metabolised at both urban and bodily scales. It has been discussed how these actions somehow restore their right to be metabolised, accentuating how capitalist forces in the urban sphere undermine their own nature, even if it is assimilated by labourers or the environment. In general, therefore, it seems that promoting the right to metabolism, either individually and communally, is essential to achieve a conciliation between urban spaces and nature-society relations in the benefit of people. These spatial, physiological and psychological needs are recognised in the work provided by Alessandra Lai in Chapter "Temporary Appropriation in Shanghai and Hong Kong: Two Study Cases Assessing the Resilience of Women Faced with the Lack of Affordable Housing", describing how female workers struggle to take part in the construction of the urban social landscape in the city.

The possibility to take part in the urban process in the city is not exclusive of TA in public spaces, TA seems to occur in the informal settlements, such as the case of informal settlements explained by Brendan Josey and Diego Ramirez-Lovering in Chapter "(Temporary) Appropriation (of Space), Makassar, and Urban Kampung". It explains how TA could help to evidence formerly hidden social practices, meanings, and flaws of ongoing development patterns, and thus be leveraged and operationalised to inform the design of community engagement practices and built environment solutions.

The insights gained from this study may be of assistance to architects, urban designers and planners because the spatial configuration of a place is recognised as a triggering factor of TA and social interaction, through the definition of values, behaviour and actions (Martínez, 2014), a key to a successful and positive transformation of the built environment is its resilience, understood as being able to best withstand the changing environment and society; such as New Zealand's case explained by Paola Boarin in Chapter "Extending Temporary Appropriation Through Architecture: The Role of Adaptive Reuse in Shaping New Zealand's Built Environment". Moreover, it has been set the relevance of TA for understanding the social landscape of the urban environment by looking at activities happening in public spaces as discussed in Chapter "Using the Street in Mexico City Centre: Temporary Appropriation of Public Space Versus Legislation Governing Street Use". It stressed the versatile nature and function of public spaces, illustrating how they are used in traditional and unexpected ways in a creative manner. While the latter emerges in a somehow organic manner, it also occurs the opposite when urban designers, planners, architects and activist help residents to intervene in public spaces as described by Luisa Bravo in Chapter "Transforming Everyday Public Space: Human Appropriations in Search for Citizenship and Urban Well-Being", encouraging TA based on creative and participatory practices.

The findings from this study make several contributions to the current literature. First, TA hints the imaginary, symbols and rituals that people's value have towards a specific place. Second, it expresses an interaction between the individuals (or collectively) and their city through specific activities happening in the urban landscape, re-affirming their right to fully use and manage their everyday life in the city. Third, it emerges in spaces where people feel identified. Lastly, TA relies on the relationships and flows product of the interaction between other landscapes within the urban environment such as socio-cultural, urban design, legal framework, economic and environmental. Being limited to TA of public space, this overall research does not provide insights of the phenomenon at larger urban scales. A natural progression of this work would be to focus on TA on different urban scales such as neighbourhoods, districts and cities, and look at the implications of TA in relation to the political or criminal dimension of the city. Further research might explore deeper the relationship between TA and relevant topics such as mental health and well-being in urban areas.

References

Fonseca Rodriguez, J. M. (2015). La importancia y la apropiación de los espacios públicos en las ciudades. *Revista de Tecnologia y Sociedad, 4*(7), 1–11.

Gehl, J. (2011). *Life between buildings*. Washington, DC: Island Press.

Jacobs, J. (1961). *The death and life of great American cities* [Online]. Available from: https://doi.org/10.2307/794509.

Martínez, E. (2014). Configuracion urbana, hábitat y apropiación del espacio. *Scripta Nova. Revista Electrónica de Geografía y Ciencias Sociales, 493*(33), 1–20.

Minnery, J. (2012). Sidewalks: Conflict and negotiation over public space. *Urban Policy and Research, 1146*, 1–2 [Online]. Available from: https://doi.org/10.1080/08111146.2012.729694.

Why Can't We Build Cities That Make Us Feel Safe, Happy and Healthy Anymore?

Marini, Riccardo

We evolved as a very social animal, and we created settlements that facilitated that trait. The types of places we created provided the basic needs of shelter and security, and this enabled human flourishing. We evolved sophisticated ways of living in close proximity to each other developing technologies to facilitate this. This proximity facilitated trade commerce and sociopolitical development. In the nineteenth century, we became so good at being clever animals to the extent that we began to dismantle the things that made our cities good to be in. We have a legacy of nearly 100 years of corrupting, with wonderful theories, the habitats we had created. The notion of public space is at the core of a good place, a space that equalizes the social differences we have instituted. On average, between 25 and 37% of our cities are made up of streets which are the base building block of public space. We are creating cities not based on the right elements or requirements because we have codified the wrong things, and the nature of the public spaces we create is critical to the type of city we want. This is not rocket science; it is simple and it is place making.

Man the Walking Animal

We are very social animals, and we evolved in large nomadic family groups. It is always useful to try and understand why things are the way they are, but the world is full of theories to justify the wrong things. Basic starting point for me is what makes us happy, safe and feel good. You may ask why? It is very simple, and the part of the world I have been fortunate to be born in and where my early imprinting comes from (Pistoia, Italy) has been nearly at the top of Maslow's hierarchy of needs since way before my birth.

What becomes important when you are privileged to live in such a state is totally different than places where you have to worry about the basics of life's survival. This brings with it many complex issues related to the sociopolitical mechanisms when have constructed that itself would need more than a book to unravel. I want to focus

A. Melis et al. (eds.), *Temporary Appropriation in Cities*,
https://doi.org/10.1007/978-3-030-32120-8

on the notion of public space, its creation, its appropriation, its use, its misuse, its degradation and the total misreading of it and its reasons for being.

I have been very fortunate in my professional life that part of what I have been asked to do is to think about cities, and it became clear to me soon after I left university that cities are a reflection of the culture of the peoples that created them. Without that, to me obvious, consideration, it is difficult to work out the necessary understanding of public space.

So, evolution of culture is critical to understanding the nature of the places that we have evolved.

> 'first we shape places then they shape us' Jan Gehl

Approximately six million years ago we became bipedal, we started to walk! Simply, before the curse of the seed that gave us the freedom to begin to create the efficient world we live in, we found that we could go and chase down a wild boar, having the upper limbs free, and we could carry it back home to feed the extended family. This protein intake initiated the development of our brains. We are at the beginning of culture, we started to sing, make music, dance, and we then developed language and writing. This is all linked to our ability to walk, we became walking animals!

> "…an action that is simultaneously an act of perception and of creativity of reading and writing the territory" Francesco Careri

As Francesco Careri asserts, walking was also at the beginning of us altering and then shaping our territory. Imagine, as humans started to move around they needed to remember where they came from, so first, maybe they broke a tree branch to remember, but it was not permanent, so they started to change the landscape in a more perpetual way, thus they made tracks. The beginning of our streets.

Streets the Basic Building Blocks of a Place

The afore mentioned streets lead to places where we live, meet and exchange. The dimensions of such streets followed anthropometric needs and what humans were able to carry. Where two tracks intersected the chances of meeting someone else increased, the possibility of trading or of finding a mate grew so cross roads became important and the space usually swelled to create space to accommodate more people.

Jan Gehl often uses this '*maður er manns gaman*'—'man is man's greatest joy' which is an old Norse saying; people watching is one of our species favourite pass times, and I am sure it is intrinsically linked to the evolution of public space. As settlements were developed, cross-roads became nodal points, changing their nature, becoming into a street. A street is a place where people have a reason to linger, this dragging effect had an impact over the shape and settlement growth.

There are many types of settlements, but they all evolve from an opportunity to live in safety exchange and meet. I love Pompei, and I think it is my favourite dead place! By today's standards, it was a small place, and we know that size matters. It had a degree of sophistication and simplicity, and the core to it was the basic building block of all good places, the street. Roads lead you to Pompei but once you crossed the gates, they became streets, streets that where imbued with the human scale. I believe that they had traffic issues like all significant settlements would have had. However, even at faster speed than a chariot could go people on both sides of the street had the possibility to engage with each other.

The street and its nature are at the core of a good place, and the street is the primary public place.

The Curse of the Seed

We discovered farming, and for most of us the nomadic existence stopped. This change has had many repercussions for us in the twenty-first century. We can trace many of the positive changes to our sociopolitical structures to the evolution of settlements. The tribal/territorial struggles made us ingenious in how we killed and defended ourselves, and medicine philosophy and technology are in a way all based on the growth of our places. Hand in hand with this are the seeds of our destruction, change in diet, change in the levels of activity or movement we were involved in. It is only know that one can track back and begin to see where and how deeply rooted the evolution of place and the changes that accompanied/facilitated it are linked to our well-being.

Being very clever, we started to develop efficient approaches to our settlements, and with the advent of the Industrial Revolution our places started to change and intensify. The size of the logistical components of place started to bring strains to the type of settlements we had evolved. People like Frederick Taylor brought a new dimension to our lives and our places. The birth of the Industrial Revolutions goes hand in hand with the notion of efficiency. Taylor successfully linked time and money, and the driver for the evolution of our places had changed. At this time, the creation of the internal combustion engine started to impact cities, and the shift from the human scale had begun. We started to change our places to facilitate one mode of movement. Fitz Lang's Metropolis (1929) is a fantastic window into the then believed future; modernity had arrived. Pause and think, time = money, today we are conditioned to comply with this. This is not good for our mental health or well-being; if you want to sit and watch a sunset, it is costing you money!

Our Very Own Extinction Event

Fast forward to the twenty-first century, ouch!

Most of our great cities started life before the advent of cars, but they still functioned when they came along. Very quickly, our streets were being turned into roads, because this is what efficiency indicated was the right thing to do.

The streets and squares were distorted to "make cars happy", the paraphernalia needed to keep motor vehicles moving efficiently, thus polluting the visual realm and corrupting the nature of the public spaces. One of my bet-noir, the extreme symptom of this change is the introduction of roundabouts in city centers, this is a anti human device purely designed to keep flows fluid and make life difficult or nigh impossible for pedestrians and cyclists but let's face it reduces the problemed caused by drivers jumping traffic lights! So, we can justify their uncontrolled proliferation, lazy thinking.

With progress came great wealth and opportunities, and this was accompanied by high levels of pollution caused by industry and the heating of our buildings. We soon realized that industries had to be controlled and made clean, and on the whole we have been rather successful at dealing with that issue. But we are still struggling with something that we find very addictive cars! And they are killing us in many different ways.

Simply put, we changed our habitat to make cars happy, this creates very toxic environments for us, and if that was not enough, we have blindly altered our cities to make it easier for cars to circulate which means it is really difficult to walk, cycle and crucially linger.

This is a wonderful cocktail of self-destruction, pollution that is insidious; we can refuse adulterated food, we can avoid polluted water, but we have not developed the ability to not breath poisoned air (and most of the times the poison is inodorous and invisible!) and we are consigned to be stuck in interminable traffic jams or to have long commutes so our mental health is suffering, our public spaces are not for us anymore and they are noisy, polluted and uninviting. De facto, our cities are telling us not to walk or cycle but to jump into an Uber or a private car as this is what people should do.

Polluted air, noisy places, streets and squares jammed with cars = inactivity, which one of the biggest killers we have. We are the only species in the history of the planet that is orchestrating its own personalized extinction event, we are doomed!

There Is Light at the End of the Tunnel! Its Only About Invitations, but the Choice Is Yours

Well we are very clever animals, but remember we are walking animals. When you analyse historic places, you will always find a reason for the public spaces, even if they are now only marginal traces, pageantry religious or military, commerce, exchange, rest and sustenance. We could say the space was appropriated for a specific use or I would assert it is not as clear as that spaces are created even if without a clear plan. Anyway, most cities have historic cores that are blessed with space that can easily be reverted to being people centric. When you look at modern, i.e., places created after the 1947 (year the planning act was introduced in the UK), what you usually find is the Architect, Master planner or Planner playing with patterns and trying to ensure that cars move efficiently around. Not a recipe for the kind of place I want to have a coffee in.

We have codified the way we should behave in our cities. Please try to make a reflection about that and you will realised that the rules we have created are focused on how to make the place efficient and economically successful. Well my measure for a successful place is one that attracts and retains people, LIFE!

For the avoidance of any doubt, without life you do not have trade, commerce or economy!

The answer is to ensure that our places have meaning beyond car parking and efficient logistics and invite us to linger, walk and cycle.

The latter is remarkable because it is an incredible virtuous cycle, by designing places where people are at the core of the conceptual thinking we reduce the negative impact of the cars such as their toxicity and addictiveness, ensuring people's potentiality of being active and meet with others without possible risk.

So, it's your choice, demand more of your politicians, what kind of city do you want, that should be the question you are asked, a city of motorways or of human scaled streets, a city where your children and grandparents can safely, comfortably perambulate and linger. Appropriate public space by curating it and using it as we evolved it a place to meet and exchange ideas and goods a place to be happy in.

Afterword

Oren Lieberman

If something is appropriated it is 'taken as one's own'. But here I am referring not to financial ownership, which is, after all, a concept in that money's value is only true in a relative sense as accessed through symbolic language. I am not saying that the monetary notion of ownership is unreal, or without tremendous impact in our lives. I am only proffering a different understanding of what we might rightly say is 'ours'. For me, what is 'ours' refers to that which is inseparable from our bodies, i.e. what we call the environment with which we are in constant intersection, constant overlap, indeed, from which we cannot be differentiated. So, we do not occupy a space as separate from us; in a sense, we are 'spacing', creating the stuff of our experience as we go with many close kin, space being one of our closest. Take walking, as an example. It is impossible to separate the pavement upon which we are walking from the act of walking and therefore from our bodies: We are in this sense one thing: walking. This is where appropriation taking as one's own really means acknowledging not a separate self-occupying, but a seamless ecology of stuff in intimate relationship.

With temporary appropriation, these two understandings entwine: People appropriate corporate/state-'owned' city spaces as their own in order to more consciously and fully realize their natural (though usually veiled and subverted) merged identity with space. While ownership in terms of an inseparability of bodies and environments is the essential ground of our being, the discourse of ownership in terms of corporate and state power is a strong one, one that usually overpowers the former understanding. And it is this one which helps us move towards our true nature in inseparability, as it fundamentally questions our rights as citizens, inciting us to appropriate spaces which we, through this discourse, see as not our own. Often, these moves are temporary, of short duration. Yet they can be effective in shifting our own relationship to those appropriated spaces and therefor, seen as examples, to the city as a whole. Often it is indeed the short, sharp occupation of space as a kind of incision into the politico-corporate bodies which can produce shifts in our sense of ownership, having effects beyond the actual occupation. At the same time, the notion of temporary may be used to describe durations of appropriations within and across various daily rhythms. For instance, the appropriation of space in some cities by people cooking and selling 'street' food occurs not across days but at temporary times during daily rhythms.

The notion of temporary appropriation is useful in describing, and therefor eliciting, various engagements, actions and movements through and with 'space' which question normed and what are, or are interpreted to be, dictated behaviours and flows, and which remind us of our real co-ownership of the world through our bodies' inseparability from space. This book provides a range of perspectives on this temporary appropriation, from its relationship to informality, to understanding it as an assemblage within other configurations; from its powerful positioning as an 'invisible' force through duration rather than as construction, to its usefulness in crime prevention; from its 'stickiness' as day-labour hiring sites, to homelessness; from developing design strategies through its marking of space, to understanding the potential to de-temporalize appropriation through adaptive reuse; from its relationship to the 'spirit of the law' regarding street use, to the resilience of female communities; from citizens' rights, to identification with place. The breadth presented here is laudable; at the same time, the depth of each journey into a specific aspect of temporary appropriation coalesces enquiry into a substantial contribution. This selection of contributions significantly extends the very contemporary debate on cities and citizens.

Glossary of Key Terms

Autoridad del Centro Histórico An institution that supports the activities of the Head of Government, where the powers of the Public Administration Units of the Government of Mexico City (CDMX) are concentrated by each delegation. It acts under the criteria of unity, autonomy, functionality, efficiency, coordination and impartiality for the planning and ordering of the territorial, economic and social development of the Historic Centre

Autoridad del Espacio Público An institution in charge of strategically designing, planning and developing public spaces in Mexico City (Gobierno de la Ciudad de Mexico, 2016)

Built environment Man-made structures, features, and facilities viewed collectively as an environment in which people live and work (Oxford University Press, 2015)

Cultural heritage Monuments: architectural works, works of monumental sculpture and painting, elements or structures of an archaeological nature, inscriptions, cave dwellings and combinations of features, which are of Outstanding Universal Value from the point of view of history, art or science; – groups of buildings: groups of separate or connected buildings which, because of their architecture, their homogeneity or their place in the landscape, are of Outstanding Universal Value from the point of view of history, art or science; – sites: works of man or the combined works of nature and of man, and areas including archaeological sites which are of Outstanding Universal Value from the historical, aesthetic, ethnological or anthropological points of view (UNESCO, 2015)

Cultural landscapes This refers to cultural properties and represents the "combined works of nature and of man" designated in Article 1 of the Convention. They are illustrative of the evolution of human society and settlement over time, under the influence of the physical constraints and/or opportunities presented by their natural environment and of successive social, economic and cultural forces, both external and internal (UNESCO, 2015)

Fideicomiso Centro Histórico of Mexico City An institution which depends on the government of Mexico City. Among their main objectives are the promotion, management and coordination of actions and services between private investors

A. Melis et al. (eds.), *Temporary Appropriation in Cities*,
https://doi.org/10.1007/978-3-030-32120-8

and public authorities that protect, recover and preserve MCC. Also, to design specific actions and projects for the benefit of MCC (Gobierno de la Ciudad de Mexico, 2016)

Mexico City Centre The biggest city centre in the American continent; it is located at the core of Mexico City, it compounded by 668 blocks with a surface of 9.7 km^2. In 1987 it was declared by the UNESCO (2013) as a cultural heritage site due to its uniqueness, authenticity and integrity of the built-environment and culture (CONACULTA, 2017)

Musealisation It is a current strategy for transforming urban spaces, and exerts significant influence on our social, cultural and aesthetic efforts directed towards visible reconstruction of the past (Muller, 1999)

Normalisation A term originating in statistics. It eliminates the unit of measurement by transforming the data into new scores with a mean of 0 and a standard deviation of 1. These transformed scores are known as Z-scores (Abdi & Williams, 2010)

Pedestrianisation The action or process of prohibiting wheeled or motor traffic from a road, esp. in the centre of an urban area, in order to allow its free use by pedestrians (Oxford University Press, 2015)

Sacralise To endow with sacred significance (frequently through ritual); to set apart from ordinary life or use as sacred (Oxford University Press, 2015)

Sedibility An element of the streetscape refering to the number of seating opportunities visible in each picture such as benches, low walls or café chairs (Porta & Renne, 2005)

Social sustainability Within the urban context this aims to not compromise the social needs that the built environment can satisfy for future generations. It is a combination of designing both physical and social realm, infrastructure to support social and cultural life (Woodcraft, 2015)

Streetscape The spatial arrangement and visual appearance of built and landscape features when viewed from the street (Anon, 1979)

Temporary appropriation (TA) The act in which people use public spaces to carry out individual or collective activities other than the purpose that space was originally designed for (Fonseca Rodriguez, 2015)

Urban regeneration Comprehensive and integrated vision and action which leads to the resolution of urban problems and which seeks to bring about a lasting improvement in the economic, physical, social and environmental condition of an area that has been subject to change (Roberts, 2016: p. 17)

The manufacturer's authorised representative in the EU is Springer Nature Customer Service Centre GmbH, Europaplatz 3, 69115 Heidelberg, Germany. If you have any concerns regarding our products, please contact ProductSafety@springernature.com

Printed and bound by CPI Group (UK) Ltd, Croydon, CR0 4YY
30/09/2026
02225687-0002